THE COMPLETE
IDIOT'S
GUIDE® TO

Algebra Practice Problems

by Jane P. Gardner

ALPHA

A member of Penguin Group (USA) Inc.

This book is dedicated to my parents, Roger and Joan Parks. My father spent countless hours reviewing and re-explaining some of these very math concepts to me at the kitchen table. And my mother fostered in me a general love of learning.

ALPHA BOOKS

Published by the Penguin Group

Penguin Group (USA) Inc., 375 Hudson Street, New York, New York 10014, USA

Penguin Group (Canada), 90 Eglinton Avenue East, Suite 700, Toronto, Ontario M4P 2Y3, Canada (a division of Pearson Penguin Canada Inc.)

Penguin Books Ltd., 80 Strand, London WC2R 0RL, England

Penguin Ireland, 25 St. Stephen's Green, Dublin 2, Ireland (a division of Penguin Books Ltd.)

Penguin Group (Australia), 250 Camberwell Road, Camberwell, Victoria 3124, Australia (a division of Pearson Australia Group Pty. Ltd.)

Penguin Books India Pvt. Ltd., 11 Community Centre, Panchsheel Park, New Delhi—110 017, India

Penguin Group (NZ), 67 Apollo Drive, Rosedale, North Shore, Auckland 1311, New Zealand (a division of Pearson New Zealand Ltd.)

Penguin Books (South Africa) (Pty.) Ltd., 24 Sturdee Avenue, Rosebank, Johannesburg 2196, South Africa

Penguin Books Ltd., Registered Offices: 80 Strand, London WC2R 0RL, England

International Standard Book Number: 978-1-61564-091-1

Library of Congress Catalog Card Number: 2010919341

13   12   11   8   7   6   5   4   3   2   1

Interpretation of the printing code: The rightmost number of the first series of numbers is the year of the book's printing; the rightmost number of the second series of numbers is the number of the book's printing. For example, a printing code of 11-1 shows that the first printing occurred in 2011.

Printed in the United States of America

Note: This publication contains the opinions and ideas of its author. It is intended to provide helpful and informative material on the subject matter covered. It is sold with the understanding that the author and publisher are not engaged in rendering professional services in the book. If the reader requires personal assistance or advice, a competent professional should be consulted.

The author and publisher specifically disclaim any responsibility for any liability, loss, or risk, personal or otherwise, which is incurred as a consequence, directly or indirectly, of the use and application of any of the contents of this book.

Most Alpha books are available at special quantity discounts for bulk purchases for sales promotions, premiums, fundraising, or educational use. Special books, or book excerpts, can also be created to fit specific needs.

For details, write: Special Markets, Alpha Books, 375 Hudson Street, New York, NY 10014.

Publisher: *Marie Butler-Knight*

Associate Publisher/ Acquiring Editor: *Mike Sanders*

Executive Managing Editor: *Billy Fields*

Development Editor: *Ginny Bess Munroe*

Senior Production Editor: *Janette Lynn*

Copy Editor: *Tricia Liebig*

Cover Designer: *William Thomas*

Book Designers: *William Thomas, Rebecca Batchelor*

Indexer: *Brad Herriman*

Layout: *Rebecca Batchelor, Ayanna Lacey*

Senior Proofreader: *Laura Caddell*

Contents

Introduction

I had my own struggles with math as a high school and college student. I can remember sitting in high school algebra and being more focused on the fact that my teacher had left several curlers in her hair that morning than on learning the quadratic equation or how to graph inequalities. I can remember working through math problem after math problem, sheets of paper with half-solved problems and crossed-out mistakes littering my bedroom floor. Funny enough, a look at my home office over the past few months would find the same thing—any blank space on to-be-recycled office paper and envelopes were covered with algebra problems and calculations. Just goes to show that you never really know what the future holds for you.

I have grown to look at algebra, and math in general, as a puzzle. I can remember being overwhelmed with getting the right answer and worrying about completing a problem. Many years of math, and probably many years of just life, has allowed me to step back and look at algebra as a process. A series of steps that have a certain logic to them. My personal approach to math today is not so much as a mechanical exercise, as it once was, but as a problem-solving, logical exercise. And this has come from practice.

Not everyone approaches math as an art that must be practiced. We all know that good sports players practice, musicians spend hours each day practicing their instrument, and artists perfect their craft over years and years. Math is the same way.

This book is intended to provide a lot of practice. There are examples that have been worked out in step-by-step details. There are helpful hints to make things easier and common pitfalls to avoid. There are more problems that allow you to practice what you have learned. And finally there are problems that put the concepts you are learning into a concrete, applicable form. It is user-friendly math.

I suggest working through the book from the beginning. Math, like so much in life, builds on itself. There is certain logic to the order in which the material is presented. It may be a little different from the way others may do it, but I think you will find that things flow along. Start at the beginning and be ready to be challenged and also shown that hey, maybe this isn't so bad after all.

What You'll Find in This Book

This book is presented in seven parts:

In **Part 1, Warming Up with a Little Math,** you will cover many of the mathematical principles that are needed to be successful in algebra. You explore absolute values, fractions, scientific notation, and the order of operations.

In **Part 2, Equations and Inequalities—Solutions and Graphs,** you are ready to jump into the real algebra material. You'll solve one-, two-, and multi-step equations and work through several ways to graph them. You then move on to solving and graphing inequalities.

In **Part 3, Solution Sets and Matrix Math,** you continue for a bit with equality and inequalities as you explore solving and graphing sets of solutions. Then you turn to the matrix, where you add, subtract, multiply, and divide sets of numbers arranged in a matrix.

Part 4, An Introduction to Polynomials, Radicals, and Functions, becomes a little more complicated as you work to factor and solve polynomials. Then you move on to simplifying radical expressions and solving radical equations. You also look at linear functions and learn to graph them and work with special functions such as inverse and square root functions.

In **Part 5, Quadratic Everything,** you work with a part of algebra that just won't go away. That is the quadratic equation. Here, you learn to graph quadratic functions, solve quadratic equations in a variety of ways, and work with quadratic inequalities.

Fractions are back in **Part 6, Can We Please Just Be Rational Here?** Only this time the fractions are more complicated and complex as you work to solve and graph the solutions to rational equations.

Finally, in **Part 7, Odds and Ends, Data and Words,** you encounter some topics in algebra that have yet to be covered. You solve problems involving probability and odds, learn the different ways to visually represent data, and tackle word problems that may actually have an impact on your life. Finally, algebra becomes real!

Help Along the Way

Although I have tried to present the material here clearly and succinctly, there are times when a little extra help might be in order. This book also contains tips, warnings, and interesting bits and pieces of information. You'll find four kinds of sidebars:

DEFINITION

Algebra is a mathematical discipline. But that doesn't mean that there are not a ton of terms and words associated with it. Throughout this book, you will find definitions to remind or inform you about the specific terms that are being used. Everyone needs a little help once in a while.

PSST—TRY IT THIS WAY

Have you ever thought, "There must be an easier way"? There quite often is an easier way. These sidebars provide you with helpful hints or alternative ways of looking at a problem. The goal is for you to be successful—not for you to take the most difficult path.

LOOK OUT!

You are going to make mistakes (that is what erasers are for). And believe me, you won't be the first to make them. These sidebars will help point out places where others (including me) have fallen victim to mistakes or misconceptions.

REALITY CHECK

"Really, when will I ever use this stuff?" That is not an uncommon phrase to hear from students of algebra. These sidebars give examples of how the algebra you are learning can be applied to real life.

Acknowledgments

Thanks to my editor Mike Sanders. The opportunity to work on this book has been one of the highlights of my professional writing career. I have learned a lot under his guidance.

Thanks to development editor Ginny Munroe, senior production editor Janette Lynn, and copy editor Tricia Liebig.

Thanks also to my agent, Bob Diforio, for his support and for helping make this process smooth from start to finish.

And a very special message to Eric. His continued support and encouragement, with this book and with everything else in our shared lives, has helped make it possible for me to continue to grow and learn.

Special Thanks to the Technical Reviewer

The Complete Idiot's Guide to Algebra Practice Problems was reviewed by an expert who double-checked the accuracy of what you'll learn here, to help us ensure that this book gives you everything you need to know about algebra. Special thanks are extended to Gene McCormick.

Trademarks

All terms mentioned in this book that are known to be or are suspected of being trademarks or service marks have been appropriately capitalized. Alpha Books and Penguin Group (USA) Inc. cannot attest to the accuracy of this information. Use of a term in this book should not be regarded as affecting the validity of any trademark or service mark.

Warming Up with a Little Math

Ready to go? You have come this far so let's get started. There is some review you need to get through before *really* starting algebra. Think of it like warming up before your piano recital, the big game, or the first act. It's time to get thinking about math and the examples and problems here should do the trick.

Kitchen Sink Mathematics

In This Chapter

- Evaluating expressions using variables
- Interpreting and using absolute values
- Determining the order to group parts of an expression
- Understanding mathematical properties

Have you ever heard of the expression "everything but the kitchen sink"? You might say that you had everything but the kitchen sink packed into your locker or in a suitcase as you pack for a trip. Obviously, no one has a kitchen sink in their locker, but that's the point. A kitchen sink is difficult to carry into a school and even more difficult to shove into a locker. There is a lot that could be stored in your locker, or a suitcase, or a book on algebra, before you'd even consider adding a kitchen sink to the mixture. This is how you can think of this chapter. I've included a lot of math here that might seem like it is coming at you fast and furiously. But, don't worry; I stopped short of adding the kitchen sink!

Say Cheese! Smiles and Other Expressions

Algebra is the language of numbers and letters. When speaking the language of mathematicians and algebra students, the letters are called *variables*. Algebraic expressions have at least one variable.

> **DEFINITION**
>
> **Variables** are letters used in mathematics to represent one or more numbers. In the expression $y = 4x$, y and x are both variables.

Example 1

Evaluate each expression when the value of $x = 7$.

What you want to do here is to plug in the number 7 every time you see an x in each of the expressions that follow:

$27 \times x$

> This is equal to 27 multiplied by 7, which is, tah dah, 189.

$\dfrac{112}{x}$

Using the value you have been given, this expression now reads $\frac{112}{7}$, which is equal to 16.

$x - 35$

Plug in 7 for the x and you have the equation $7 - 35 = -28$.

$77 + x$

Once again, replace the variable x with the number 7 to get $77 + 7$, which is equal to 84.

Now It's Your Turn

Try your hand at some problems now. The answers can be found in Appendix A at the end of the book. No peeking until you are finished though!

Evaluate the expressions that follow when $z = 5$.

1. $8z$

2. $\frac{35}{z}$

3. $68 - z$

4. $z + 9$

Evaluate these expressions for $y = -7$.

5. $9y$

6. $\frac{y}{14}$

7. $57 - y$

8. $10 + y$

Absolute Values

Absolute values are always positive. An absolute value of a number is the distance between that number and 0 on a number line. The absolute value of a is written as $|a|$. Because absolute values are always positive, $|a| = a$ and $|-a| = a$.

Example 2

$$|y| = \frac{1}{3}$$

In this example, the value of y could be either $\frac{1}{3}$ or $-\frac{1}{3}$. The absolute value of y is $\frac{1}{3}$. The lines around the y are the absolute value sign.

Example 3

What is the absolute value of –19? Keep in mind that you might see it written as $\left|-19\right|$. The absolute value of a negative number is the positive of that number, so …

$$\left|-19\right| = 19$$

Now It's Your Turn

Solve each of the following problems. The answers can be found in Appendix A.

What is the absolute value of each of the following?

9. $\left|8 - 4\right|$

10. $\left|4 - 8\right|$

11. $\left|7 - 4 + (-6)\right|$

12. $\left|-6 - 9 - 1\right|$

13. $\left|-12 + 6 - 6 + (-8)\right|$

14. $\left|100 + (-25) - 25\right|$

Grouping

Did you flip through this book when you bought it? Did you happen to notice that there are a lot of parentheses and brackets in the equations? There are a few simple rules to follow for interpreting these correctly, and it all comes down to grouping.

There are certain symbols that give hints as to how an equation or an expression should be grouped. The next example walks you through some of the options.

PSST—TRY IT THIS WAY

When there are a series of parentheses, brackets, or absolute value signs, work from the innermost grouping first. Start there and work your way out to the edges of the expression.

Example 4

Simplify the following:

$$27 \div \{9 - 6\}$$

Start inside the brackets. If you subtract those two numbers, the expression then becomes

$$27 \div 3 = 9$$

Example 5

Look at another one: $\left|1 - 4 - (-2)\right|$.

Those absolute values are back. Start inside the absolute value symbol. Look for double negatives and remove them. $-(-2)$ is equal to $+2$, so now you have $\left|1 - 4 + 2\right|$.

Solve this from left to right. That would be "one minus four plus two" if you were to write it out. $\left|-1\right|$

Now do the absolute value magic. The absolute value of -1 is 1 so $\left|1 - 4 - (-2)\right| = 1$.

> **LOOK OUT!**
>
> When you see the symbol for absolute value, you might be tempted to just get rid of the negative signs. After all, the absolute value of a negative number is positive. But wait! Solve what is inside the absolute value symbol first, and *then* see what is left. That is when you can perform the "absolute value magic" and turn negative numbers into positive ones.

Now It's Your Turn

Try a few more examples on your own. When you are finished, you can find the answers in Appendix A. Simplify each of the following mathematical expressions:

15. $7(13 - 5)$

16. $16 \div (2 + 2)$

17. $4(9 + 1) - (4 + 5)6$

18. $\left|-6 - (5 + 4)\right|$

19. $\left|6 - (5 + 4)\right|$

20. $(4 - 5)3$

21. $-8 + (6 \times 2) - 5$

22. $[15 \div (4 + 1)] - 3$

Mathematical Properties

There are certain "rules" that math equations or statements follow. You need to know these.

The Associative Property

The associative property is based on the fact that the way things are grouped does not matter. That is, it doesn't matter for addition or multiplication.

The associative property for an addition problem says the way you group three numbers that are added together does not change the sum:

$$(a + b) + c = a + (b + c)$$

In multiplication, the way you group three numbers that are being multiplied does not change the product:

$$(f \times g) \times h = f \times (g \times h)$$

Example 6

Solve the equation 9 + (3 + 1).

If you add the numbers in the parentheses first you are left with the equation 9 + 4, which is equal to 13. It is also fine to think of this as (9 + 3) + 1 because 9 + 3 = 12 and when you add 1 to that you get the same answer, 13. This all works because of the associative property.

The associative property for multiplication is similar. It indicates that the way that three numbers in a product are grouped does not change the product. For example, $(a \times b) \times c = a \times (b \times c)$.

Example 7

Let's use the associative property with multiplication:

$$8(2 \times 7)$$

If you again start with the numbers inside the parentheses, you are left with the equation 8(14), which is equal to 112. But if you were to write the equation so that it reads (8 × 2)7, you still get the answer 112. Works out pretty well, doesn't it?

The Commutative Property

Order doesn't matter with the commutative property—when you are multiplying or adding, that is.

The commutative property of addition says that the order which two numbers are added does not matter. $a + b$ is the same as $b + a$.

And, you guessed it, the same is basically true for the commutative property of multiplication. The order in which two numbers are multiplied does not impact the product. That is because $x \times y$ is the same thing as $y \times x$.

Example 8

$m + n = n + m$

$4 + (-3) = -3 + 4$

No matter which way you write it, the answer is always 1. The same is true for multiplication:

$m \times n = n \times m$

In other words:

$2 \times 4 = 8$

This is the same as

$4 \times 2 = 8$

PSST—TRY IT THIS WAY

At first glance, the associative property and the commutative property seem somewhat similar. So think of it this way—the associative property is all about who (or what) you associate with. Numbers can group together in different small groups but the answer will still be the same (if you are multiplying or adding). The commutative property just changes the order. The order of the numbers can be totally different but as long as the numbers are the same (and you are multiplying or adding) then the answer will be the same!

LOOK OUT!

The commutative property does not work for subtraction:

$8 - 3 - 1 = 4$, but $3 - 8 - 1 = -6$

These are not the same!

The Identity Property

If you want to keep a number the same, use the identity property. Simply add 0 to a number or multiply a number by 1. This keeps the number the same by use of the identity property. The identity property for addition says that the sum of any number and 0 is equal to the original number. The identity property for multiplication says that the product of a number and 1 is equal to that number.

Example 9

$$a + 0 - a$$

Or, in other words:

$$9 + 0 = 9$$

$$b \times 1 = b$$

or

$$8 \times 1 = 8$$

Inverse Properties

There is a lot of "cancelling" that happens in algebra. Cancelling is when you can eliminate some part of an equation. For example, suppose you were asked to solve the following equation:

$$1 + 2 - 2 + 5 =$$

You might notice that there is a positive 2 and a negative 2 in the equation. These cancel each other out. You could just cross them out so you were left with only $1 + 5$ which is 6.

Sometimes when adding, you need to use the identity property. This says that the sum of a number and its opposite is 0. In addition, 0 is called the additive identity element. The use of the identity property and the additive identity element helps make cancelling in addition problems much easier.

When cancelling in multiplication, you want to get back to the multiplicative inverse element (or 1). So you need to multiply a number by its *reciprocal* to get back to 1. Check out the examples that follow to see how this is done (don't worry, it is a lot easier than it might sound at first).

DEFINITION

The **reciprocal** of a number is one divided by that number. In other words, the reciprocal of 3 is $\frac{1}{3}$. The reciprocal of 89 is $\frac{1}{89}$, and the reciprocal of 7483 is, you guessed it, $\frac{1}{7483}$.

Example 10

Use the inverse property to find the additive identity element of 19:

$$19 + (-19) = 0$$

In this example, you started with the number 19. You added the opposite to it to come up with an answer of 0. 0 is the additive identity element. Use the reciprocal of –2 to find the multiplicative inverse element:

$$-2 \times (-\frac{1}{2}) = 1$$

Here you started with –2. You then multiplied this number by its reciprocal, $-\frac{1}{2}$. As a result, you were left with the multiplicative inverse element or 1.

Now It's Your Turn

Identify the mathematical property that makes each of the following statements true.

Remember, you can check your answers when you are done by turning to Appendix A.

23. $6 \times 8 \times 2 = 2 \times 8 \times 6$

27. $(1 \times 2) \times 3 = 1 \times (2 \times 3)$

24. $14 \times \frac{1}{14} = 1$

28. $28 + 77 + 123 = 77 + 123 + 28$

25. $-99 + 0 = -99$

29. $85 + (-85) = 0$

26. $(4 + 7) + 3 = 4 + (7 + 3)$

30. $-\frac{1}{7} \times 1 = -\frac{1}{7}$

Fractions, Fractions, Everywhere

In This Chapter

- Simplifying fractions
- Using the least common denominator to add and subtract
- Multiplying fractions
- Dividing fractions

Remember your friend the fraction? Two numbers or variables placed one on top of the other with a line between them? If the used record store is having a sale on old CDs at 20 percent off, then they cost $\frac{1}{5}$ of what they usually would. Tell someone you will be done in 45 minutes and you actually mean it will take you an additional $\frac{3}{4}$ of an hour. If your share of an 8-slice pizza is 3 slices, then you actually get to have $\frac{3}{8}$ of the pizza. Fractions are everywhere. You use them in your everyday life all the time. This chapter helps you manipulate and use fractions to the full extent of their usefulness.

Simplifying Fractions

Sometimes, fractions are in a form that is not quite finished. There is more that you could do to the fraction. And usually, that means that the fraction needs to be simplified. Teachers (and algebra authors) like to see fractions in their simplified form. It makes them think that you really are paying attention. Here's how you do it.

Example 1

$$\frac{10}{15}$$

What on Earth do you do with *that* number? Hang on, we'll work it out.

In this case, the number 10 is the numerator of the fraction, and 15 is the denominator. Many times in algebra, fractions need to be in their simplest form. That is, the numerator and denominator have no common factors except for 1.

$\dfrac{10}{15}$ is not in the simplest form. Both 10 and 15 can be divided evenly by 5, right? Therefore, 5 is the *greatest common factor*. So divide the numerator and denominator by 5 like so:

$$\frac{10}{15} = \frac{10 \div 5}{15 \div 5} = \frac{2}{3}$$

DEFINITION

The **greatest common factor** or two or more whole numbers (that are not zeros) is the largest common factor that they have. This would be the largest whole number that is a factor of each.

So the simplified version of $\dfrac{10}{15}$ is $\dfrac{2}{3}$. Let's try another one.

Example 2

Write the fraction $\dfrac{60}{96}$ in the simplest form.

If you take a quick look at this fraction, you might see that 6 will divide into the numerator and denominator evenly. In a rush to get this problem solved, and move onto the next one, you do just that—you divide the 60 by 6 and the 96 by 6.

$$\frac{60 \div 6}{96 \div 6} = \frac{10}{16}$$

Now, take a minute to look at your answer. Is $\dfrac{10}{16}$ in the simplest form? No, both of those numbers are divisible by 2. Keep going with this fraction to get the true simplest form:

$$\frac{10 \div 2}{16 \div 2} = \frac{5}{8}$$

Done! $\dfrac{5}{8}$ is in the simplest form. Keep in mind that you do not have to "hit" the greatest common factor the first time you attempt a problem. But make sure you keep going until you are truly in the simplest form. Ok, try one more before going out on your own.

Example 3

Write the fraction $\dfrac{77}{88}$ in the simplest form.

At first glance, it might appear as though these numbers have little in common. But actually each can be divided evenly by 11 (in other words 11 is the greatest common factor). So …

$$\frac{77 \div 11}{88 \div 11} = \frac{7}{8}$$

You are ready to try some of these on your own.

Now It's Your Turn

Write each of the following fractions in the simplest form. When you are done, turn to Appendix A to check your answers.

1. $\dfrac{40}{48}$

2. $\dfrac{4}{12}$

3. $\dfrac{7}{49}$

4. $\dfrac{5}{43}$

5. $\dfrac{54}{81}$

6. $\dfrac{24}{64}$

7. $\dfrac{22}{24}$

8. $\dfrac{72}{12}$

9. $\dfrac{3}{9}$

10. $\dfrac{68}{68}$

Finding the Least Common Denominator

Simplifying fractions is one thing. Being able to add them or subtract them is a whole other story. This process, which is covered next, requires that you do one necessary step. You must find the *least common denominator* of a set of fractions.

DEFINITION

The **least common denominator** (also known as LCD to its friends) is the least common multiple of the denominators of two or more fractions.

Example 4

Find the least common denominator of these two fractions and then add them:

$$\frac{3}{5}+\frac{1}{6}$$

The least common denominator of these two fractions is 30. Sure, they also have numbers such as 60, 120, and 240 in common, but to keep life simple, use the *least* common denominator.

Now each fraction has to be changed into a form that uses 30 as the denominator.

In the case of $\frac{3}{5}$, the denominator would need to be multiplied by 6 to equal 30. So multiply both the numerator and denominator by 6. Now $\frac{3}{5}=\frac{3\times 6}{5\times 6}=\frac{18}{30}$.

In the case of $\frac{1}{6}$, the denominator would need to be multiplied by 5 to equal 30. So multiply both the numerator and denominator by 5 in this case: $\frac{1}{6}=\frac{1\times 5}{6\times 5}=\frac{5}{30}$.

Now the equation reads

$$\frac{18}{30}+\frac{5}{30}$$

These fractions can now be added together. The answer is $\frac{23}{30}$.

PSST—TRY IT THIS WAY

So you really want to find a common denominator of two numbers? Take the short cut and simply multiply the two numbers together. That can tell you a common denominator very quickly!

Example 5

Find the least common denominator for these two numbers and then subtract

$$8-\frac{1}{2}$$

Remember that the number 8 does have a denominator; it is 1. The least common denominator of these two fractions is 2:

$$\frac{8\times 2}{1\times 2}=\frac{16}{2}\qquad\qquad\frac{16}{2}-\frac{1}{2}=\frac{15}{2}$$

If you are not completely happy with the number $\frac{15}{2}$, then you can change it to $7\frac{1}{2}$. It's the same thing!

Now It's Your Turn

Add or subtract each of the following, finding the lowest common denominator when necessary. You can find the answers in Appendix A.

11. $\dfrac{3}{9}+\dfrac{4}{9}$

16. $\dfrac{8}{9}+\dfrac{1}{3}$

12. $\dfrac{3}{7}-\dfrac{3}{2}$

17. $\dfrac{2}{3}-\dfrac{1}{6}$

13. $\dfrac{1}{8}+\dfrac{4}{5}$

18. $\dfrac{6}{9}-\dfrac{3}{8}$

14. $\dfrac{3}{5}-\dfrac{5}{3}$

19. $\dfrac{5}{6}+2$

15. $\dfrac{7}{4}-3$

20. $\dfrac{6}{5}-\dfrac{5}{6}$

Operations on Fractions

You just worked through problems where you had to add and subtract fractions. Grab a scalpel. We are going to multiply and divide them now. They are just numbers, remember, and they can't feel a thing.

Multiplying Fractions

Multiplying fractions is as easy as multiplying two numbers. You multiply the numerators together and then multiply the denominators. And then you are done!

Example 6

Multiply $\dfrac{7}{5} \times \dfrac{2}{3}$

Multiply the two numbers in the numerator together. And then multiply the two numbers in the denominator together. It looks like this:

$$\frac{7 \times 2}{5 \times 3} = \frac{14}{15}$$

And you are done!

> **PSST—TRY IT THIS WAY**
>
> Having trouble remembering exactly what you are supposed to do here? You might want to keep the product rule in mind. The product rule can be summed up with this equation: $\dfrac{a}{b} \times \dfrac{x}{y} = \dfrac{ax}{by}$.

Try another one.

Example 7

Multiply $\dfrac{1}{9} \times 9$

Multiply the two numbers in the numerator and then the two numbers in the denominator.

In this example, that would be

$$\frac{1}{9} \times 9$$

In most cases, your teacher (or friendly algebra author) would suggest simplifying that fraction. $\dfrac{9}{9}$ is just a fancy way of writing 1. The answer to this example is 1.

> **LOOK OUT!**
>
> Remember that a single number has a numerator and a denominator. A whole number has a denominator of 1. It is quite possible that fact will need to be used once in a while so don't forget it!

Dividing Fractions

Are you a fan of division? Not really? Well, in this case, that is fine because dividing fractions is a lot like multiplying fractions, except there is a little flipping that has to go on.

When fractions are divided, they are actually multiplied by the *reciprocal* of the fraction.

Take a look at this example.

Example 8

Divide $\dfrac{4}{5} \div \dfrac{3}{2}$

The reciprocal of $\dfrac{3}{2}$ is $\dfrac{2}{3}$. That is the number you want to use here!

Now the equation is a multiplication equation:

$$\frac{4}{5} \times \frac{2}{3} = \frac{4 \times 2}{5 \times 3} = \frac{8}{15}$$

So ...

$$\frac{4}{5} \div \frac{3}{2} = \frac{8}{15}$$

> **DEFINITION**
>
> Two nonzero numbers whose product is equal to 1 are **reciprocals**. The reciprocal of $\dfrac{2}{3}$ is $\dfrac{3}{2}$ because $\dfrac{2}{3} \times \dfrac{3}{2} = 1$.

Example 9

Divide $\dfrac{1}{8} \div \dfrac{4}{7}$

Flip the second fraction (that is, find its reciprocal) and then multiply:

$$\frac{1}{8} \div \frac{4}{7} = \frac{1}{8} \times \frac{7}{4} = \frac{1 \times 7}{8 \times 4} = \frac{7}{32}$$

Now It's Your Turn

Multiply or divide each equation. Be sure to check your answers in Appendix A.

21. $\dfrac{5}{6} \div \dfrac{6}{5}$

22. $\dfrac{3}{4} \times \dfrac{8}{9}$

23. $\dfrac{1}{7} \times \dfrac{2}{4}$

24. $\dfrac{3}{6} \div \dfrac{3}{2}$

25. $\dfrac{3}{8} \times \dfrac{1}{2}$

28. $\dfrac{5}{9} \times \dfrac{4}{8}$

26. $\dfrac{4}{5} \div 6$

29. $\dfrac{1}{2} \times \dfrac{1}{2}$

27. $\dfrac{6}{7} \div \dfrac{6}{7}$

30. $\dfrac{2}{6} \times \dfrac{2}{7}$

Math Expressions 101

In This Chapter

- Working with exponents
- Using scientific notation
- Using the order of operations

Are you starting to feel warmed up? A few chapters of variables, fractions, and absolute values and you are starting to get comfortable, I hope. Let's continue with our introduction, or possibly review, of some of the algebra basics you need to work through the chapters to come.

Exponents

You've got the power. In math that is. Mathematically speaking, a power is any expression that represents the repeated multiplication of the same factor. For example, 1,024 is a power of 4. If you were to multiply 4 five different times as in 4 × 4 × 4 × 4 × 4, you would get the number 1,024.

Now, it's a bit cumbersome isn't it, writing 4 × 4 × 4 × 4 × 4 every time you want to show that 4 multiplied five times is equal to 1,024 (admit it, you can see yourself doing that many times each day). So algebra experts have come up with a way to write the power in a different way. A power can be written as a base and an exponent.

In this case, 4 is the base and 5 is the exponent. That tells you how many times the base is multiplied together 4^5.

Exponents can be positive or they can be negative. The positive ones are fairly straightforward to work with. The negative ones are a little more complicated.

Example 1

Here is an example of a problem that involves a negative exponent.

A negative exponent looks like this:

$$7^{-3}$$

This is not what it appears to be. 7^{-3} is the same thing as $\frac{1}{7^3}$. See what happened here? The negative sign in the exponent was a signal for you to put the entire number in the denominator of a fraction. This is known as the *reciprocal*.

> **DEFINITION**
>
> **Reciprocals** are two numbers whose product is 1. For example, 3 and $\frac{1}{3}$ are reciprocals because when they are multiplied, the answer is 1. The reciprocal of a number with a negative exponent, such as a^{-n}, is equal to $\frac{1}{a^n}$.

Now you can solve it. In this case, $7 \times 7 \times 7 = 343$. So $7^{-3} = \frac{1}{343}$.

Example 2

What if there is an exponent of 0, as in 8^0?

Here is an easy one. Any number raised to the zero power is equal to 1. So $8^0 = 1$!

> **LOOK OUT!**
>
> Raising a number to the zero power is not the same as having zero raised to a power. 0^0 is undefined and is also undefined for negative exponents. 0 to any positive exponent is equal to 0. If you were to write the formula for negative or zero exponents in equation form,
>
> they would be $a^0 = 1$ when $a \neq 0$ and $a^{-n} = \frac{1}{a^n}$ when $a \neq 0$!

Multiplying Exponents

When multiplying exponents, you add the exponents.

Example 3

Find the product of the following powers:

$$7^6 \times 7^3$$

If you were to multiply this all out it would look like this

$$(7 \times 7 \times 7 \times 7 \times 7 \times 7) \times (7 \times 7 \times 7)$$

Count them, there are nine 7s here. That means that 7 has been multiplied 9 times. That can be written as 7^9. The same answer you would have gotten if you had simply added the exponents in the original equation.

> **REALITY CHECK**
>
> If you have ever used a microscope in science class, you probably have had to multiply exponents. The total magnification that you get under the microscope is the product of the objective lens and the eyepiece. In some microscopes, the objective lens magnifies an object 10^2 times while the eyepiece magnifies it 10^1 times. Suddenly it is an algebra problem, not a science problem!

Dividing Exponents

If you divide two numbers with exponents, the exponents are subtracted.

Example 4

Simplify the following expression:

$$\frac{5^4}{5^2} = 5^2$$

The exponents 4 and 2 were subtracted to get the exponent of 2.

Now It's Your Turn

Evaluate each expression. You can find the answers in Appendix A.

1. 10^2

2. 9^3

3. 5^4

4. $\left(\frac{2}{5}\right)^4$

5. 87^0

6. $\frac{4^6}{4^1}$

7. $(-5)^{-3}$

Find the product or quotient of each of the following.

8. $5^6 \times 8^0$

9. $\frac{4^6}{4^1}$

10. $\dfrac{7^4}{7^1}$ 11. $1^2 \times 1^8$

Scientific Notation

Remember Pluto? The distant neighbor to Neptune, which was once listed as the ninth planet in our solar system? Pluto was stripped of its status as a planet in 2006 and is now listed as a dwarf planet because it was too small to be considered a planet any more. Poor Pluto. Pluto is in orbit around the Sun at an average distance of three billion, six hundred seventy three million, five hundred thousand miles. Writing that out in numbers is 3,673,500,000. That is one big number. There must be an easier way to write this number and there is. You can write it using scientific notation, which uses exponents.

Example 5

To write a number like that in scientific notation, you move the decimal place (there is a decimal place to the right of the last zero 3,673,500,000.00) until you are to the left of the last number. So in this case, you would move the decimal point to the right of the leading 3, like so

    3.673500000

How many spaces did you move it? Count as you go.

Nine spaces. So you can now rewrite the distance between Pluto and the Sun as 3.6735×10^9. The exponent here is the number of spaces you moved the decimal point to the left! You keep all the other nonzero numbers as part of the number when it is in scientific notation. That helps to accurately represent the distance from the Sun to Pluto.

Just as there are very big numbers, such as needed for distances in space, there are also very small numbers.

Example 6

Did you know that a 1 attometer is equal to 0.000000000000000001 meters? Written in scientific notation that would be 1.0×10^{-18} m. If you count, you would have to move the decimal point 18 places to the right to get it to the right of the 1 in this example.

PSST—TRY IT THIS WAY

If the line of zeroes is in front of the number, you move the decimal to the right. This results in a negative exponent. If the line of zeros is behind the number, you move the decimal to the left. This results in a positive exponent.

Now It's Your Turn

Write the following numbers in scientific notation. The answers are located in Appendix A.

12. 975,000

14. 57,580,000

13. 0.00000478

15. 0.010101

Write the following numbers in standard form.

16. 6×10^{-3}

18. 9.0048×10^{8}

17. 4×10^{6}

19. 5.487×10^{-4}

Order of Operation

If you are going to make scrambled eggs, you have to follow some basic steps in a certain order. It is usually helpful to turn the stovetop burner on before trying to cook the eggs. If you were going to add butter or oil to the pan to prevent sticking, then you should do that before adding the eggs. And most importantly, crack the egg shell before dropping the egg into the pan.

Algebra is not unlike scrambled eggs. There is a certain order in which things have to happen in algebra to ensure that you come up with the correct answer to a problem. And to ensure that you don't eat too many egg shells.

There are four steps that you must take to solve expressions that have more than one operation:

1. Evaluate inside the parentheses or brackets.

2. Evaluate powers and exponents.

3. Multiply and divide from left to right.

4. Add and subtract from left to right.

Follow these four steps and you'll be on your way to successful algebra problem solving.

PSST—TRY IT THIS WAY

Many people use mnemonics to remember the order of operations. One common one is Please Excuse My Dear Aunt Sally (P—parentheses, E—exponents, M—multiplication, D—division, A—addition, S—subtraction). You may want to remember that or make up your own. What about Purple Elephants Make Delicious Apple Scones?

Example 7

$$64 \div 2^3 + 3 - 5$$

Okay, there are no parentheses here so you can start with step 2.

Solve for the exponent. In this case it is 2^3. This is equal to 8. Substitute that into the equation. It is now:

$$64 \div 8 + 3 - 5$$

Step 3 says to work from left to right by multiplying and dividing. If you do this, the expression now looks like …

$$8 + 3 - 5$$

Now proceed to step 4. Add and subtract from left to right:

$$11 - 5 = 6$$

Example 8

$$67 - (4^2 + 1)$$

Remember, step 1 says to evaluate what is inside the parentheses or brackets. And it just so happens there is an exponent inside. So work on the exponents first:

$$67 - (16 + 1)$$

$$67 - 17$$

Now subtract from left to right:

$$67 - 17 = 50$$

Now It's Your Turn

Evaluate the following expressions. You can find the answers in Appendix A.

20. $3(5 + 9)$

21. $5[(7 - 3) \div 4]$

22. $17 \times 6^2 - 8$

23. $\frac{1}{4}(4 + 8) - 6^2$

24. $4[10 - (5 - 3)^2]$

25. $8^2 \div 4 + 16$

Evaluate each expression when $x = 6$.

26. $x^2 - 7$

27. $15 + x - 9$

28. $3(x^2 + 3)$

29. $\frac{x - 6}{3x + 8}$

30. $3x \div 2$

Equations and Inequalities—Solutions and Graphs

Okay, here we are. On the brink of *real* algebra. You are very comfortable with basic equations and inequalities: $5 + 5 = 10$. $6 + 4 \neq 12$. $6 > 4$. All good stuff. The next things we will be doing will look very similar but you will find that at least one x, y, or other variable will be scattered in the problem somewhere. As you work along, you will find that algebraic equations and inequalities are not all that different from the equations you can already solve.

Ready for Basic Training— Equation Training, That Is

In This Chapter

- Solving one- and two-step equations
- Solving multi-step equations
- Understanding absolute value equations and their two solutions
- Solving problems involving ratios, proportions, and percents

Ready for some real algebra? The past three chapters of this book have been what I would call an over-all review of math with a little bit of algebra thrown in to keep us honest. Now it's time to get down to the basics and start on algebra. So, buckle up, it's going to be a bumpy ride.

Solving One-Step and Two-Step Problems

The most important thing to remember when solving a one-step problem is that you want to isolate the variable. In plain English this means you want to rearrange the equation so that the variable is all by itself. After you do that, the equation is easy to solve. I promise.

Example 1

Solve the equation $y + 8 = 14$.

You want to get y all alone on one side of the equation. The best (and possibly the only) way to do that is to subtract 8 from that side. But if you subtract 8 from one side you have to subtract 8 from the other side, too. You have to be fair here: $y + 8 - 8 = 14 - 8$

Now the equation looks like this: $y = 6$

What you used here was the *subtraction property of equality* to solve this problem. As you might guess, there is also an *addition property of equality*. See the following example.

Example 2

Solve for z in the following:

$$z - 6 = 12$$

To solve this equation, you need to get the variable z all alone. That can be done by adding 6 to both sides of the equation. Now you have …

$$z - 6 + 6 = 12 + 6$$

$$z = 18$$

It is possible to multiply and divide in a one-step problem as well.

Example 3

Solve for d in the following:

$$-7d = 63$$

To get the variable on its own, you need to divide the left side of the equation by –7. Because you are doing that to one side of the equation, you have to divide the right side of the equation by –7, too:

$$\frac{-7d}{-7} = \frac{63}{-7}$$

$$d = -9$$

Sometimes you have equations that take two steps to solve. It's just like the examples you just walked through, but with one more step (which is why they are two-step problems).

Example 4

Solve this two-step equation:

$$\frac{k}{3} - 3 = 21$$

The goal is the same here. You want to get the variable, k, alone on one side of the equation. The first thing to do is to get rid of that –3 there. Do that by adding 3, to both sides of the equation:

$$\frac{k}{3} - 3 + 3 = 21 + 3$$

$$\frac{k}{3} = 24$$

To finish this up and get k alone, 3 needs to be multiplied to each side of the equation:

$$\frac{k}{3} \times 3 = 24 \times 3$$

$$k = 72$$

Now It's Your Turn

Solve the following equations. You can find the answers in Appendix A.

1. $y + 7 = 34$

4. $\frac{y}{6} - 5 = 37$

2. $12 - t = 16$

5. $14.4a - 3.5 = 3.7$

3. $-8x = 56$

6. $13 + 8y = 53$

Solve each word problem by writing the equation and solving it.

7. Winson has a small garden in his backyard. The total area of the garden is 180 square feet. The width of the garden is 15 feet. What is the length?

8. Jenn buys three yards of felt to make scarves for her friends. Each scarf takes 0.75 yards of cloth. How many scarves can she make with her felt?

9. A local music store offers piano lessons. They charge $18 per class and a registration fee of $10. Emmett paid a total of $712 to take lessons in one year. How many lessons did he take?

10. An ice skating rink charges $9 per session to skate and an additional $3 per session to rent skates. If someone goes to the ice rink 15 times in three months and spends a total of $165, how many times did they forget their skates and need to rent them?

Solving Multi-Step Equations

You've solved one- and two-step equations. All that is left now are the multi-step equations.

Example 5

Solve this equation:

$$6x - 2x + 12 = 32$$

The first thing you might notice is that there are two numbers with the variable x. The best way to start this is to combine like terms (in other words, add them together).

Now the equation looks like this: $4x + 12 = 32$

This looks like one of those two-step equations from the last section. Treat it that way. Subtract 12 from each side:

$$4x + 12 - 12 = 32 - 12$$

$$4x = 20$$

Divide each side by 4:

$$\frac{4x}{4} = \frac{20}{4}$$

$$x = 5$$

Example 6

Try this one. Solve this equation:

$$6y + 3(y + 5) = 60$$

See those parentheses there? Use the distributive property to get rid of them. Now the equation looks like this:

$$6y + 3y + 15 = 60$$

Combine the like terms:

$$9y + 15 = 60$$

Solve for *y:*

$$9y + 15 - 15 = 60 - 15$$

$$9y = 45$$

$$\frac{9y}{9} = \frac{45}{9}$$

$$y = 5$$

LOOK OUT!

When using the distributive property, it is important to distribute the number outside the parentheses to all the numbers inside the parentheses. Sometimes people tend to just distribute it to the first number.

Example 7

One more multi-step equation and you'll be ready to solve problems on your own!

$$\frac{3}{4}(7x+6) = 15$$

There is that fraction hanging out there in the front. To remove that from the left side of the equation, you need to multiply it by its reciprocal. And if you multiply the left side of the equation by the reciprocal of $\frac{3}{4}$ then you need to multiply the right side of the equation by the reciprocal of $\frac{3}{4}$.

$$\frac{4}{3} \times \frac{3}{4}(7x+6) = 15 \times \frac{4}{3}$$

Now it looks like this:

$$7x + 6 = \frac{60}{3} = 20$$

Subtract 6 from both sides to start to isolate the variable:

$$7x + 6 - 6 = 20 - 6$$

$$7x = 14$$

Divide both sides by 7 and you are done:

$$\frac{7x}{7} = \frac{14}{7}$$

$$x = 2$$

Now It's Your Turn

Solve the following multi-step equations. The answers are located in Appendix A.

11. $5x + 4(x - 2) = 15$

15. $-6 = 2b - 24 - 8b$

12. $\frac{2}{5}(x + 6) = 12$

16. $16 = \frac{2}{7}(8 - 6x)$

13. $7 + 4(x + 7) = 80$

17. $7w + 2(9 - 3w) = -5$

14. $14y + 18 + 10y = 24$

18. $10t + 2(t + 2) = 46$

Following are two solutions that have errors. Describe each error and correctly solve the equation.

19. $\frac{1}{6}(4y + 10) = 12$

$4y + 10 = 2$

$y = -8$

20. $s + 2s - 3 = 6$

$4s - 3 = 6$

$4s = 3$

$s = \frac{3}{4}$

Special Equations

Some equations have a little extra. For example, they may have variables and absolute value signs. Let's take a closer look at some of those special cases now.

Example 8

Solving an equation with absolute values. Solve $|t| = 6$.

t can be equal to 6 or t could be equal to −6.

LOOK OUT!

The symbol $|a|$ shows the distance between a and zero on the number line. That could be a units more than 0 (and thus a positive number) or a units less than 0 (and therefore a negative number). This is why there are two possible answers when solving an equation with an absolute value.

Example 9

Here is another equation with an absolute value for you to solve:

$$|y + 8| = 21$$

This one absolute value equation can be written as two different equations.

$y + 8 = 21$ or $y + 8 = -21$

Solve them both:

$y + 8 = 21$ or $y + 8 = -21$

$y + 8 - 8 = 21 - 8$ $y + 8 - 8 = -21 - 8$

$y = 13$ $y = -29$

The solutions to this problem are 13 and −29.

Ratios and Percents

Other special equations are *ratios* and *percents*. Let's start the discussion with ratios.

Writing Ratios

Ratios can be written in three ways: a to b, $a{:}b$, or $\dfrac{a}{b}$.

No matter which way it is written, it is always read the same way "the ratio of a to b."

> **DEFINITION**
>
> A **ratio** compares two quantities using division.
>
> A **percent** is a fraction that has a denominator of 100. For example, 23% can be written as twenty-three out of one hundred or $\frac{23}{100}$. In decimal form, 23% is 0.23 or twenty-three hundredths.

Example 10

There are 12 girls in Ms. Niang's classroom and 15 boys. What is the ratio of girls to boys? And of boys to girls? There are two things you are looking for here. First is the ratio of girls to boys:

$$\frac{girls}{boys} = \frac{12}{15} = \frac{4}{5}$$

Second is the ratio of boys to girls:

$$\frac{boys}{girls} = \frac{15}{12} = \frac{5}{4}$$

Another special equation that is useful in many situations is the proportion.

Example 11

The generalized formula for a *proportion* is $\frac{a}{b} = \frac{c}{d}$ where $b \neq 0$ and $d \neq 0$. Solve this proportion:

$$\frac{12}{3} = \frac{x}{4}$$

Once again, you want to get the variable x alone on one side of the equation. In this case, that can be done by multiplying both sides of the equation by 4:

$$4 \times \frac{12}{3} = \frac{x}{4} \times 4$$

$$\frac{48}{3} = x$$

$$x = 16$$

> **DEFINITION**
>
> A **proportion** is an equation that indicates that two ratios are equivalent. You read a proportion as "a is to b as c is to d."

Percents

One of the most useful applications of a proportion is when you are talking about percents. Percent problems can be solved using a proportion. Percents can be expressed as proportions. If you have the statement "*a* is *x* percent of *y*," then the proportion you would set up is as follows:

$$\frac{a}{y} = \frac{x}{100}$$

PSST—TRY IT THIS WAY

A long time ago, when I was learning about percents, I learned a trick that you might find useful. Most percent problems have the words "is" and "of" in them, as in what *is* 25 percent *of* 56? Or 12 *is* 76 percent *of* what number. Use this handy formula to plug the values into and then solve for the variable: $\frac{is}{of} = \frac{\%}{100}$. It works every time.

Example 12

What percent of 30 is 15? Write the proportion:

$$\frac{a}{y} = \frac{x}{100}$$

$$\frac{15}{30} = \frac{x}{100}$$

Cross-multiply:

$$15 \times 100 = 30x$$

$$1500 = 30x$$

Solve for *x* by dividing both sides by 45:

$$x = 50$$

So 30 is 50% of 15.

Now It's Your Turn

Solve the following special equations. The answers are located in Appendix A.

21. $2|7x - 9| - (-6) = -2$

22. $|y + 7| = 8$

23. $|p| = 0.5$

24. During the past 12 days, there were 4 sunny days, 2 cloudy days, and the rest were rainy. Write the proportion of rainy days to sunny days.

25. You need to add 3 cups of oatmeal to each batch of oatmeal cookies. Each batch makes 36 cookies. How many cookies can you make if you have 9 cups of oatmeal? Set up the proportion and solve the equation.

26. Alina can read 8 pages of her book in 15 minutes. How many pages will she read in 45 minutes?

27. The baseball coach is encouraged. Her team scored a total of 12 runs in the first 3 games of the season. If this trend continues, how many runs will her team score in the 10 games that are left in the season?

28. Tristan walks 0.25 miles to the bus stop each morning. If that is 10% of the total trip to his school, how far is it from his home to his school?

29. 86 is 40 percent of what number?

30. Eric wants to leave a 18% tip at a restaurant. The total bill was $58. How much of a tip should he leave?

Linear Equations and Their Graphs

In This Chapter

- Using a table to graph linear equations
- Graphing linear equations using intercepts
- Learning about the slope of a line and the rate of change

Linear equations are huge in algebra. In this chapter, you will solve a lot of problems involving linear equations and more specifically, their graphs. Grab some graph paper, a pencil, and a ruler and let's get started.

Linear equations are equations where the graph is a line (hence, the term "linear" equation—get it?). The standard form that most linear equations fit into is …

$$Ax + By = C$$

This is true for values of A, B, and C, which are real numbers as long as A and B are not both zero.

Graphing with Tables

One way to graph linear equations is with tables.

Example 1

You have been asked by your teacher, tutor, or the author of some sort of algebra practice problem best-seller to graph the following linear equation:

$$-4x + y = -6$$

The first thing you would do (after taking a deep breath and possibly a bathroom break) is to solve the equation for y. Use the skills you honed from the previous chapter of this book:

$$-4x + y = -6$$

$$y = 4x - 6$$

Keeping in mind that you need to plot a few points on the graph, the easiest thing is to make a table to record the values for x and y. Such a table needs to have values for x and y based on the equation you have been given.

Substitute numbers into the equation. Use the equation you just solved for to plug the numbers in.

If you try $x = 0$, then your equation is now $y = 4(0) - 6$ or $y = -6$.

Put that value in the table.

Try another number. Solve for $x = -1$.

Substitute into the equation $y = 4x - 6$:

$$y = 4(-1) - 6$$

$$y = -10$$

Put that in the table, too!

Try two more numbers: $x = 1$ and $x = 2$:

$$y = 4x - 6 \qquad\qquad y = 4x - 6$$

$$y = 4(1) - 6 \qquad\qquad y = 4(2) - 6$$

$$y = -2 \qquad\qquad\quad y = 2$$

Using those numbers, you should have a table that looks like this:

x	y
0	-6
-1	-10
1	-2
2	2

Now, finally, you are ready to try your hand at graphing this equation. Due to space considerations, we have created graphs with x- and y-axes that are on different scales throughout this book.

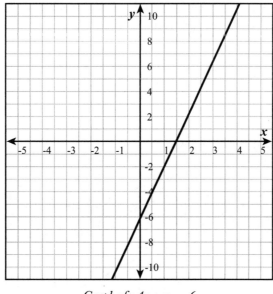

Graph of −4x + y = −6.

So that is what the line for the equation −4x + y = −6 looks like!

Example 2

Can't stop there. Let's try another one. Graph the following linear equation: $y + x = 5$. Remember, solve for y by rearranging the equation.

$$y = 5 - x$$

Now make a table to find different values for x and y. Substitute the following values for x into the equation:

$$x = 1, x = 0, x = 7, x = -4$$

And solve for y. Then put the numbers into a table, as shown here.

LOOK OUT!

Remember that any point on a graph is designated by an x value and a y value. x is on the horizontal axis and y is on the vertical axis.

x	y
1	4
0	5
7	−2
−4	9

Of course you can substitute in any values you wanted, and you can do as many as you want. But why spend more time on one problem than necessary?

PSST—TRY IT THIS WAY

Try to pick numbers that will give you positive and negative values so you have a better idea of what the line of the equation looks like.

Plot these values on a graph like the following:

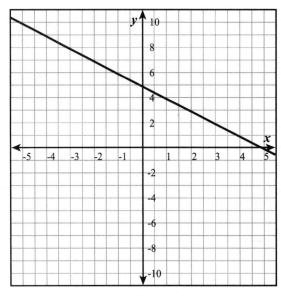

Graph of y + x = 5.

Notice anything different about these graphs? This one slopes down to the right. The first one sloped down to the left. Hmmm. Hold that thought. We'll be dealing with that later in this chapter.

Example 3

One more example and you'll be ready to work through these on your own. Making tables to find the values is not that bad, is it? Graph the equation $y = -3$.

Wait a second. There is no value for x is there? What is wrong here? Go back to the standard form for linear equations given at the beginning of this chapter:

$$Ax + By = C$$

If you recall, the caveat was that A and B could not *both* be zero. But there is no rule that says that one of them can't be zero. If you were to take the equation you are asked to graph and plug it back into the original equation, it would look like this:

$$0x + 1y = -3$$

In this case, A is equal to zero and B is equal to 1. Simplifying this, you have the new, improved equation $y = -3$. See, there is no need to panic. But how do you graph this? Make a table like the following one.

x	y
0	–3
1	–3
2	–3
–5	–3

For any value you choose for x, the $y = -3$. So an equation with just y graphs to a horizontal line.

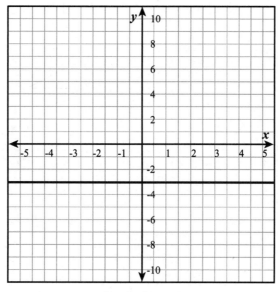

Graph of y = –3.

Now It's Your Turn

Graph the following equations. You can find the answers in Appendix A.

1. $2y + x = 4$

2. $x = 9$

3. $y - 4x = 0$

4. $3x + 4y = 8$

5. $y + 2x = 1$

7. $y - x = 0$

6. $2x + 4y = 16$

8. $y = 5.5$

Answer the following questions about linear equations and their graph.

9. What is the equation that will give the following graph?

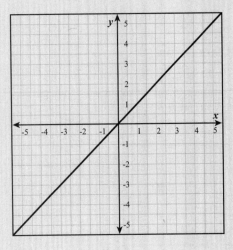

10. What is the equation that will give the following graph?

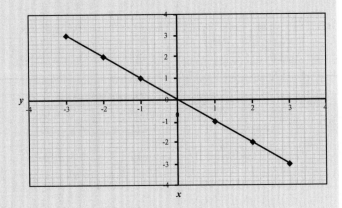

Graphing with Intercepts

Graphing by substituting numbers into a table is not the only way to graph a linear equation. A line can be drawn with only two points (similar to how you had to plot a bunch of points in the last section).

If you are going to draw a line with only two points, it helps to pick those points carefully. The best way to do that is to choose the point where the line crosses the x-axis (also known as the x-intercept) and the point where the line crosses the y-axis (also known as the y-intercept).

This section of the chapter focuses on finding the values of the x- and y-intercepts. To find the x-intercept, solve for x when $y = 0$. To find the y-intercept, solve for y when $x = 0$.

Let's see how.

Example 4

Find the x intercept and the y-intercept of the graph of $4x + 2y = 12$. Pick one to start with. If you are going to find the x-intercept, solve for x when $y = 0$. This means you plug 0 in for y in the equation. It now looks like this:

$$4x + 2(0) = 12$$

So the equation now is …

$$4x = 12$$

Solve for x by dividing both sides of the equation by 4:

$$\frac{4x}{4} = \frac{12}{4}$$

The answer is $x = 3$. This is the x-intercept.

You are only halfway done, though. You need to find the y-intercept now. Solve for y when $x = 0$. The equation now reads:

$$4(0) + 2y = 12$$

This is $2y = 12$. Divide each side by 2 to solve for y:

$$\frac{2y}{2} = \frac{12}{2}$$

Now, it is:

$$y = 6$$

So the y-intercept for the equation in question is $y = 6$.

If you were going to plot the linear equation, the points would be …

$$(0,6) \text{ and } (3,0)$$

This is because when $x = 0$, $y = 6$. This is just another way to name the point (0,6). When $x = 3$, $y = 0$ so the point is (3,0).

The graph of the equation $4x + 2y = 12$ looks like this:

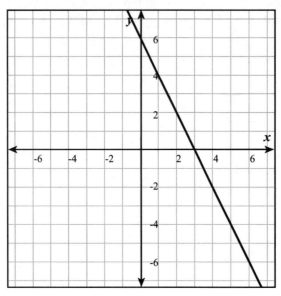

Graph of 4x + 2y = 12.

Example 5

Find the intercepts and graph the equation for $x + 3y = 9$. First, find the value of the x-intercept (remember, solve for x when $y = 0$):

$x + 3(0) = 9$

$x = 9$ (This is the x-intercept.)

Then find the y-intercept (remember, solve for y when $x = 0$):

$0 + 3y = 9$

$3y = 9$

$y = 3$ (This is the y-intercept.)

Because the x-intercept is 9, you will plot the point (9,0) and because the y-intercept is 3, you will plot the point (0,3). The graph looks like the following graph.

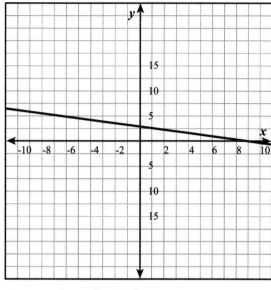

Graph of x + 3y = 9.

Example 6

You can work backward here as well. What if you were given this graph?

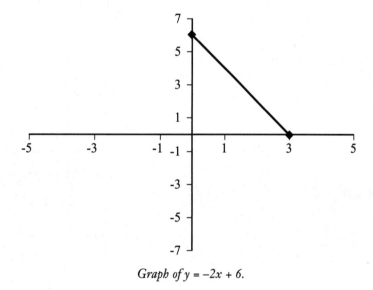

Graph of y = −2x + 6.

What Is the *x*-Intercept? What Is the *y*-Intercept?

A close inspection of the graph shows that the line crosses the *x*-axis at (3,0) and the line crosses the *y*-axis at (0,6). This line has an *x*-intercept of 3 and a *y*-intercept of 6.

Now It's Your Turn

Find the *x*- and *y*-intercepts of these graphs:

11.

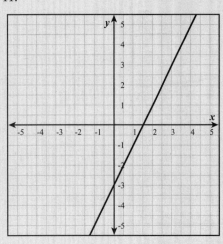

13.

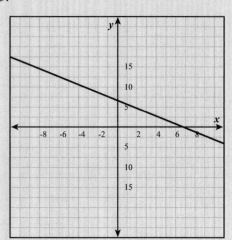

12.

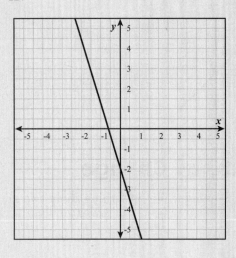

14.

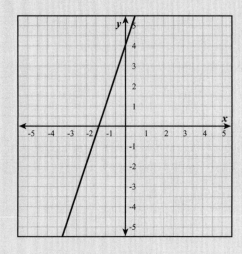

Find the intercepts and graph the following equations.

15. $7x + 6y = 42$

18. $y = \frac{1}{5}x - 10$

16. $x - y = 2$

19. $4x - 4y = 24$

17. $-15x + 3y = 5$

20. $1.5x - 0.5y = 6$

The Slope of Lines and the Rate of Change

Remember back in the first section of this chapter where I pointed out something about two graphs. One graph was slanting down to the right and one was slanting down to the left. Well, now we are going to look at that!

That observation concerns the slope of the lines. The slope of a line that is not vertical is a ratio. It is the ratio of the vertical change to the horizontal change of any two points on that line.

Another way to think of this is $\frac{\text{rise}}{\text{run}}$. The rise is the difference between the y value of two points and the run is the difference between the x values of two points.

The slope of a line is given the variable *m*.

The equation for finding the slope of a line is $m = \frac{y_2 - y_1}{x_2 - x_1}$.

PSST—TRY IT THIS WAY

Need another way to think of slope? slope $= \dfrac{\text{the change in } y}{\text{the change in } x}$.

This graph shows you how to find the slope of the line visually.

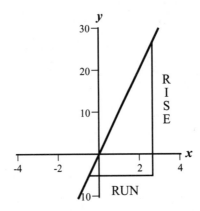

Graph of the slope of a line.

Let's put all this to work.

Example 7

Find the slope of the line on this graph.

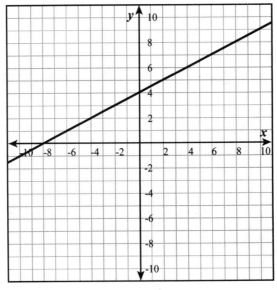

Graph of $y = \frac{1}{2}x + 4$.

Choose the point (–2,3) to be (x_1,y_1) in the equation. This makes (4,6) the values used for (x_2,y_2) in the formula.

PSST—TRY IT THIS WAY

Don't worry about which point you choose as (x_1,y_1) and which you choose as (x_2,y_2). It doesn't matter. You are choosing to help keep track of the numbers. The answer is the same no matter which you use where.

Use the formula for slope $m = \dfrac{y_2-y_1}{x_2-x_1}$ and substitute in the correct numbers:

$$m = \frac{(6-3)}{(4-(-2))} = \frac{3}{6} = \frac{1}{2}$$

The slope of the line on this graph is $\dfrac{1}{2}$! Try another one.

Example 8

What is the slope of the line that passes through the points (7,3) and (7,–3)? Set (7,3) as point (x_1,y_1) and (7,–3) as (x_2,y_2) and solve using the equation for slope:

$$m = \frac{y_2-y_1}{x_2-x_1}$$

$$m = \frac{(-3-3)}{(7-7)} = \frac{-6}{0}$$

Wait a minute! How can this be? You can't divide a number by 0. The value is undefined. Let's think about this a minute. The slope of this line is undefined. What type of line would have a slope that is undefined … a vertical line! The slope of this line is vertical.

Check it out on this graph.

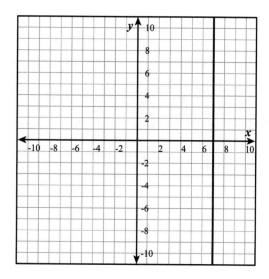

Graph of x = 7.

Different Slopes

There are different types of slopes. Some are vertical (and therefore undefined) as in the previous example. Others can be horizontal. A horizontal slope has a slope of 0.

Example 9

What is the slope of a line that has the points (–4,5) and (6,5)? Use that formula again:

$$m = \frac{y_2 - y_1}{x_2 - x_1}$$

$$m = \frac{(5-5)}{(6-(-4))} = \frac{0}{10} = 0$$

A slope of 0? Hmmm. This is a horizontal line.

Rates of Change

Fortunately, algebra is not all about numbers all the time. The whole idea of slope can be used to talk about rate of change, too.

Example 10

Sam sets up her own babysitting business. She sets up the following cost structure. Her rate for 2 hours of sitting is $7. For 4 hours she charges $14 and if she sits for 6 hours her total cost is $21.

What is the rate of change in cost with respect to time for her babysitting services?

Remember that the rate of change $= \dfrac{\text{change in cost}}{\text{change in time}}$.

Choose two of her numbers and compare them:

Rate of change $= \dfrac{21-14}{6-4} = \dfrac{7}{2} = 3.5$

The rate of change is $3.50 per hour.

Now It's Your Turn

Find the slope of a line that passes through the following points. The answers are located in Appendix A.

21. (–2,–1) and (–3,6)   22. (–4,3) and (1,4)

23. (−10,8) and (−8,2)

24. (2,3) and (6,9)

Explain why the following answers are incorrect. Then fix the error to find the correct answer.

25. The slope of the line that passes through the points (−5,−3) and (2,4).
$$m = \frac{(-5-2)}{(-3-4)} = \frac{-7}{-7} = 1$$

27. Jesse's skateboard ramp is 15 inches off the ground. The overall length of the ramp is 33 inches. What is the slope of his ramp, to the nearest inch?

26. The slope of the line that passes through the points (0,4) and (−3,1).
$$m = \frac{4-1}{-3-0} = \frac{3}{-3} = -1$$

This table shows the number of customers at a local record store during the past 10 weeks. Use this information to answer questions 28–30.

Week	Number of Customers
1	124
2	151
3	200
4	232
5	212
6	204
7	170
8	137
9	100
10	72

28. What is the rate of change between weeks 4 and 6?

29. What is the rate of change between weeks 1 and 4?

30. During which period was the rate of change the greatest? Weeks 1 through 3, weeks 4 through 6, or weeks 7 through 10?

Linear Equations, the Sequel!

In This Chapter

- Using the point-slope formula to graph a line
- Using the slope-intercept formula to graph a line
- Learning the differences between parallel and perpendicular lines and their graphs
- Using a table to graph an absolute value function

Nope, you are not done with linear equations. Remember, this is an algebra book after all. In the last chapter, you solved problems that dealt with the x- and y-intercepts, slopes, and linear equations. This chapter takes all that a bit farther.

Point-Slope Formula

We start this chapter with the point-slope formula. The point-slope formula for the equation of any line that is not vertical is …

$y - y_1 = m(x - x_1)$ where m is the slope of the line.

REALITY CHECK

This may seem like just another equation to learn. But it actually can help you a lot. After an equation is written in the point-slope form, you can easily find the x, y coordinates of a point on the line and the slope of the line.

Example 1

Suppose you have a line that passes through the point (4,3) and has a slope of $\frac{1}{2}$. Assume in this case that (x_1, y_1) is the point that you know (4,3). Here, the slope $(m) = \frac{1}{2}$.

Substitute this number into the point-slope formula:

$$y - y_1 = m(x - x_1)$$
$$y - 3 = \frac{1}{2}(x - 4)$$

That's it! The point-slope formula for that line is $y - 3 = \frac{1}{2}(x - 4)$.

Example 2

What if you had the point-slope equation? It would be fairly easy to graph that equation, right? Consider the point-slope equation $y - 5 = \frac{1}{4}(x - 3)$. Just by looking at this equation you know that the slope of the line is $\frac{1}{4}$. That is because if you compare this equation for the generalized equation for point-slope $(y - y_1) = m(x - x_1)$.

You also know one of the points here. The point (x_1, y_1) is (3,5). So what you do here is plot the point (3,5) on the graph. Using the slope of $\frac{1}{4}$, you can plot a second point. The other point is (–1,4). Connect the points and you have your graph!

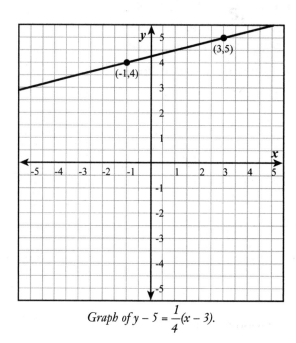

Graph of $y - 5 = \frac{1}{4}(x - 3)$.

Now It's Your Turn

Write the point-slope line equation for each of the following. You can find the answers in Appendix A.

1. (–1,4), $m = 2$

2. (–3,11), $m = -6$

3. (7,5), $m = \frac{3}{4}$

4. (–5,4), $m = -1$

5. $(3, -2)$, $m - \dfrac{2}{3}$

Graph each equation in point-slope form.

6. $y + 4 = -3(x + 2)$

9. $y - 3 = -(x - 6)$

7. $y - 5 = -2(x - 3)$

10. $y - 1 = -x + 2$

8. $y + 4 = \dfrac{2}{3}(x - 6)$

Slope-Intercept Form

A linear equation that is in the form $y = mx + b$ is in what is called slope-intercept form. In this equation, m is the slope of the line and b is the y-intercept. The examples that follow look at the use of the slope-intercept form.

Example 3

Find the slope and the y-intercept of this equation:

$y = 4x + 6$

The first one is easy. It is already in the slope-intercept form, right? It is in the format of $y = mx + b$. So in this case, the slope of the line is 4 and the y-intercept is 6. Okay, too easy. Let's try another one.

Example 4

Find the slope and the y-intercept of this equation:

$4x + 2y = 2$

This is not as clear cut, is it? The equation is not in the format of $y = mx + b$. So, you guessed it, the first thing you have to do is rearrange it so that it is in the correct form.

You need to solve for y so subtract $4x$ from each side:

$$4x + 2y = 2$$
$$-4x \quad -4x$$
$$2y = -4x + 2$$

Keep going. Divide each side by 2 so that y is all alone:

$$y = -2x + 1$$

Now it is in the slope-intercept form. This equation has a slope of –2 and a y-intercept of 1.

The slope-intercept form can also be used to graph equations. As you can tell, there is more than one way to graph an equation in algebra. So many choices …

Example 5

Graph the equation $3x + 3y = 12$. Again, this equation is not in the correct format. Rearrange so that it is in the form $y = mx + b$.

Isolate the y by subtracting $3x$ from each side:

$$3y = -3x + 12$$

Divide each side by 3:

$$y = -x + 4$$

Now you have something to work with. The slope of this line is –1 and the y-intercept is 4, which is point (0,4).

Plot that point. Use the slope to find a second point. Then connect the dots!

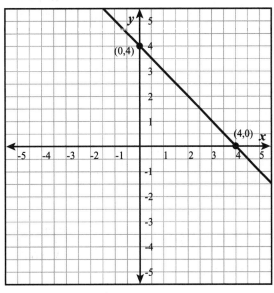

Graph of y = −x + 4.

Now It's Your Turn

Find the slope and the *y*-intercept of these equations. The answers are located in Appendix A.

11. $y = -2x$

14. $-6x + 3y = 2$

12. $3x - y = 5$

15. $x - y = -9$

13. $-x - 8y = 24$

Find the slope and *y*-intercept of the following equations and then graph the equation.

16. $y = 12 - 6x$

17. $4y = -x + 32$

18. $3y = -2x + 3$

20. $y - 3x = -6$

19. $x + y = 0$

Parallel Lines, Perpendicular Lines, and Absolute Values

Do all lines behave so nicely? Not a chance. Let's walk through examples of some of the unique lines.

Parallel Lines

Parallel lines are lines that never touch. They go on and on forever and never get any closer together nor do they get any farther apart. How is this possible?

Parallel lines have the same slope, that's how. Take a look at the example that follows.

Example 6

Determine if these two lines are parallel and then graph them:

$y = 3x + 7$ and $-3x + y = -4$

Put both equations into the slope-intercept form ($y = mx + b$). Notice that the first is in that form (hurray!).

When the second equation is rearranged, it reads $y = 3x - 4$:

$y = 3x + 7$ and $y = 3x - 4$

They both have the same slope, 3. So they are parallel (it's always good to check). Now graph them on the same graph to see what this looks like.

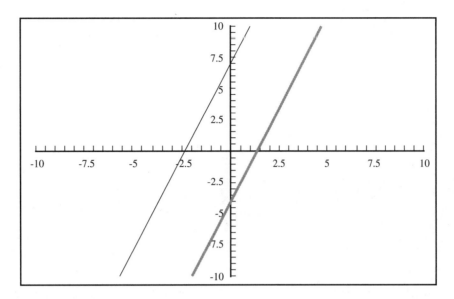

Graph of y = 3x + 7 and y = 3x – 4.

Example 7

Take a look at another example. Determine if these lines are parallel and plot them on a graph:

$$y = 3x - 4 \text{ and } 3x + y = 8$$

First, put them in the slope-intercept form:

$$y = 3x - 4 \qquad\qquad 3x + y = 8$$
$$y = -3x + 8$$

Wait a second. These lines are not parallel. The first one has a slope of 3 and the second has a slope of –3. Graph them anyway to see that they do meet and are therefore truly not parallel.

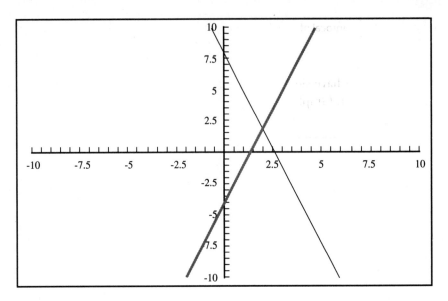

Legend

$y = 8 - 3x$

$y = 3x - 4$

Graph of y = −3x + 8 and y = 3x − 4.

Perpendicular Lines

Perpendicular lines intersect at 90 degrees (a right angle). The slopes of perpendicular lines are *opposite reciprocals*.

DEFINITION

The **opposite reciprocal** is a number by which a certain number can be multiplied that gives a product of 1. For example, the reciprocal of $\frac{2}{3}$ is $\frac{3}{2}$. The negative reciprocal of $\frac{2}{3}$ is $-\frac{3}{2}$.

Example 8

Determine if these two lines are perpendicular and then graph them:

$y = 7x + 5$ and $x + 7y = 7$

Place each equation in the slope-intercept form:

$y = 7x + 5$ and $y = -\frac{1}{7}x + 1$

Remember that the reciprocal of a whole number, a, is equal to $\frac{1}{a}$. For example the reciprocal of 196 is equal to $\frac{1}{196}$.

See how these two equations have slopes that are negative reciprocals of each other (7 and $-\frac{1}{7}$)? This means they are perpendicular. Graph them to make sure.

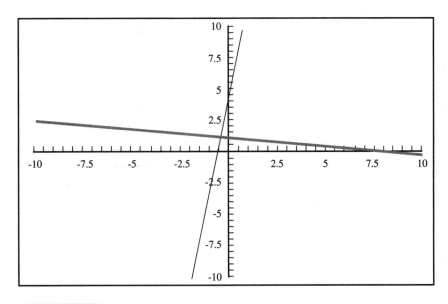

Legend

$y = 7x + 5$

$y = 1 - \frac{x}{7}$

Graph of $y = 7x + 5$ and $y = -\frac{1}{7}x + 1$.

Equations with Absolute Values

Remember absolute values? This is the distance between a number and zero on the number line.

When it comes to graphing absolute values, you need to graph the absolute value *function*. This means that you have two variables with which you need to fill in a table to get enough points to make a graph. We'll walk through a few examples to make it clearer.

A **function** has a set of inputs and a set of corresponding outputs. Each input is paired with exactly one output. These are covered in depth in Chapter 15, but for now, just keep those few facts in mind.

Example 9

Graph the absolute value of $f(x) = |x|$.

Make a table of paired inputs and outputs.

x	$f(x) = |x|$
−3	3
−2	2
−1	1
0	0
1	1
2	2
3	3

See how each value of x (input) is equal to only one value of $f(x) = |x|$ (output)?

Now graph these points to get the graph of $f(x) = |x|$.

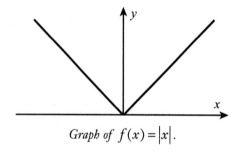

Graph of $f(x) = |x|$.

Example 10

Graph the following absolute value function:

$g(x) = -4|x|$

Make your table of values.

x	$-4|x|$
−3	−12
−2	−8
−1	−4
0	0
1	−4
2	−8
3	−12

Then graph the points.

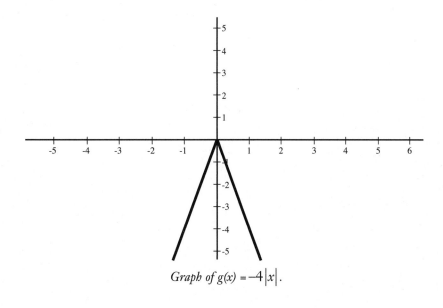

Graph of g(x) = −4|x|.

Now It's Your Turn

Graph each set of equations and indicate if the lines are parallel, perpendicular, or neither. The answers are located in Appendix A.

21. $y = x + 5$ and $4x + 2y = 4$

24. $2x + 3y = 15$ and $y = -\dfrac{2}{3}x + 5$

22. $-x + 2y = 1$ and $y = \dfrac{1}{2}x + \dfrac{1}{2}$

25. $2y - 4x = 1$ and $y = 2x + 11$

23. $y = 2x + 6$ and $y = 2x + 2$

Graph each absolute value function.

26. $f(x) = -|x|$

28. $f(x) = |x| + 4$

27. $g(x) = -3|x|$

29. $g(x) = |x| - 1$

Writing Linear Equations

In This Chapter

- Writing slope-intercept equations
- Working with point-slope equations
- Writing equations in standard form
- Writing an equation for a set of data

You've graphed 'em and talked about 'em. Now it's time to write those linear equations. This process is just like working backward on what you have already done.

Writing Slope-Intercept Equations

If you are given the slope of a line and a point on that line or a graph of a line, you can write a slope-intercept equation for that data.

PSST—TRY IT THIS WAY

When writing an equation using the slope-intercept, keep this formula in mind:

$y = mx + b$

It all comes back to that formula.

Example 1

Suppose you were given the graph on the following page.

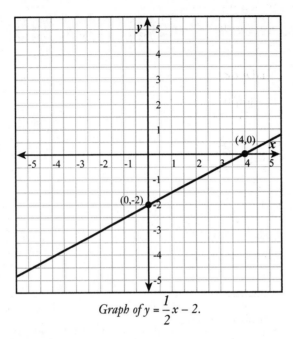

Graph of $y = \dfrac{1}{2}x - 2$.

What is the slope of this line?

Recall that the slope of a line is $\dfrac{\text{rise}}{\text{run}}$. If you use that formula, the slope of this line is equal to $\dfrac{2}{4} = \dfrac{1}{2}$.

As is seen in the graph, the line crosses the y-axis at $(0,-2)$. The y-intercept (or b) is equal to -2.

Therefore, the equation of the line in this graph is

$$y = \dfrac{1}{2}x - 2$$

Example 2

What if you were given this graph and wanted to use the two labeled points to figure out the equation of the line?

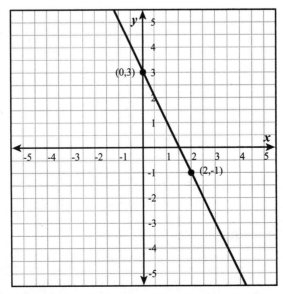

There are two points labeled on this graph: (0,3) and (2,–1).

Why not make (0,3) stand in for (x_1, y_1) and (2,–1) stand in for (x_2, y_2). If you do that, you can use the following formula to find the slope of the line:

$$m = \frac{y_2 - y_1}{x_2 - x_1} = \frac{-1 - 3}{2 - 0} = -\frac{4}{2} = -2$$

This graph has a slope of –2. The point where it crosses the y-axis is (0,3). The y-intercept is 3.

Using our favorite formula for the slope of a line, the slope of this line is

$$y = -2x + 3$$

Now It's Your Turn

Using the information given, write the equation for each of the following. Turn to Appendix A to check your answers.

1. $m = 5$, y-intercept = –3

2. $m = -1$, y-intercept = 6

3. $m = \dfrac{1}{5}$, y-intercept = 8

4. $m = -4$, y-intercept = $-\dfrac{2}{7}$

5. $m = 3$, y-intercept = 3

Use the following graphs to write the equation for the line. Use the slope of the line or the plotted points to determine your answer.

6.

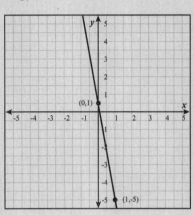

9.

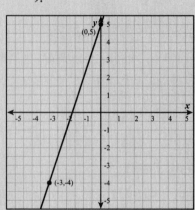

7.

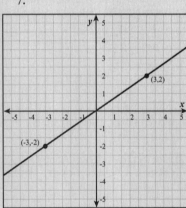

10.

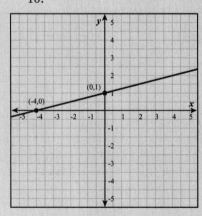

8.

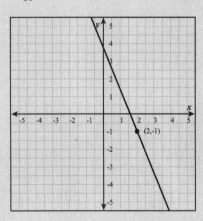

Point-Slope Equations

Moving right along. What if you were given a graph or information about one specific point on a line and its slope? With that information, you could write the equation of that line.

PSST—TRY IT THIS WAY

Remember that the point-slope form of an equation can be written as $y - y_1 = m(x - x_1)$.

Example 3

Take a look at this graph.

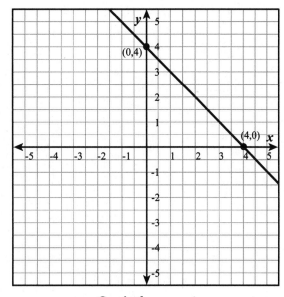

Graph of $y = -x + 4$.

Two points are clearly labeled here: (0,4) and (4,0). Use those to calculate the slope of the line using this formula:

$$m = \frac{y_2 - y_1}{x_2 - x_1} = \frac{0 - 4}{4 - 0} = -1$$

Now plug one of the identified points into the equation for the point-slope form:

$$y - y_1 = m(x - x_1)$$

$$y - 0 = (-1)(x - 4)$$

$$y = -x + 4$$

The equation for this line is $y = -x + 4$!

Good job. Why not try another one?

Example 4

Write the equation in point-slope form for a line that passes through point (–1,6) and has a slope of 2.

This is an easy one. Just plug into this formula:

$$y - y_1 = m(x - x_1)$$
$$y - 6 = 2(x - -1)$$
$$y - 6 = 2x + 2$$
$$y = 2x + 8$$

Now It's Your Turn

Use the point-slope formula to write equations for the lines that follow. The answers are located in Appendix A.

11. A line that passes through point (4,2) with a slope of –1.

12. A line that passes through point (0,–5) and has a slope of $\frac{1}{3}$.

13. A line that passes through point (–2,–5) with a slope of $-\frac{3}{5}$.

14. The line in the following graph.

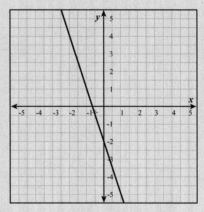

15. The line in the following graph.

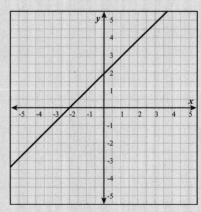

Standard Form Equations

The standard form of an equation is

$$Ax + By = C$$

All linear equations can be written in this form. In this section, our goal is to work on writing equations in that form.

Example 5

Suppose a line passes through points (1,2) and (3,–1). How do you write the equation of that line in standard form?

First thing is first. You need to find the slope of that line. You can graph the points on graph paper, but you suddenly find that your supply of graph paper is gone. Not to worry; there is a formula you can use (of course):

$$m = \frac{y_2 - y_1}{x_2 - x_1}$$

Plug in the numbers you do have, arbitrarily picking (1,2) to be (x_1,x_2) and (3,–1) as (x_2,y_2):

$$m = \frac{y_2 - y_1}{x_2 - x_1} = \frac{-1 - 2}{3 - 1} = -\frac{3}{2}$$

Write the equation in point-slope form using point (1,2):

$$y - 2 = -\frac{3}{2}(x - 1)$$

Now, rewrite what you have here in standard form:

$$y - 2 = -\frac{3}{2}(x - 1)$$

$$y = -\frac{3}{2}x + \frac{7}{2} \quad \text{Not pretty is it? But that is the answer.}$$

Example 6

You didn't think you'd have just one example did you? A line passes through point (0,12) and has a slope of $\frac{3}{4}$. How can you write the formula for that line in standard form?

Well, you have the slope and a point, so you are almost there. Write the equation using the information you have. I suggest using the point-slope formula (but hey, it's up to you how much work you want to make for yourself):

$$y - y_1 = m(x - x_1)$$
$$y - 12 = \frac{3}{4}x - 0$$

Rearrange to get this in standard form:

$$\frac{3}{4}x + y = 12$$

Tah dah!

Now It's Your Turn

Using the information given, write the equation for the line described in standard form. You can find the answers in Appendix A.

16. $m = 0$, (12,–6)

17. $m = -2$, (–4,0)

18. $m = -\frac{2}{3}$, (–4,2)

19. (–8,4), (2,–2)

20. (–1,0), (–9,–6)

Writing Equations of Parallel and Perpendicular Lines

There are always some special cases. This time, you need to look at the equations of parallel and perpendicular lines.

Parallel Lines

Railroad tracks, the opposite sides of a square, and the yellow lines in the middle of a street (providing everything goes well with the painting machine)—these are all examples of parallel lines. Parallel lines never meet and stay the same distance apart at all times.

Want to discuss parallel lines like a mathematician? Tell people that parallel lines have the same slope.

LOOK OUT!

Remember parallel lines never cross. The only way for this to happen is if they have the same slope. Slope is the measure of the rate at which a line rises or falls.

Example 7

You have one line with the equation $y = 4x + 1$. You want to write the equation for a line that is parallel to the first line and also passes through point $(3,-2)$.

The first thing you need to do is find the slope of the line you have. The slope is 4. So the line you want to find that is parallel to that also has a slope of 4.

Use the coordinates of the known point to find the y-intercept of the new line. Plug it into the equation for the slope of a line:

$$y = mx + b$$

$$-2 = 4(3) + b$$

$$b = -14$$

So the equation of the line parallel to the original line is $y = 4x - 14$. Want visual proof that these two lines are parallel? Check out this graph.

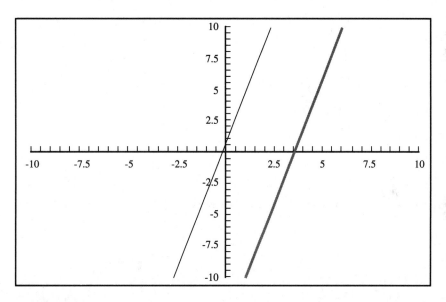

Graph of $y = 4x - 14$ and $y = 4x + 1$.

Perpendicular Lines

Two lines are perpendicular if they intersect at a right angle (at 90 degrees). A horizontal line and a vertical line on a square are perpendicular. This means that the slopes of two perpendicular lines are reciprocals.

> ✏️ **PSST—TRY IT THIS WAY**
>
> Think of it this way. Horizontal and vertical lines are perpendicular to each other.

Example 8

Suppose a line has the equation $y = -\frac{1}{2}x + 3$. What is the equation of the line that is perpendicular to that line and also passes through point (5,4)?

First thing to do is to look at the slope of the line given. The slope of that line is $-\frac{1}{2}$. Therefore, the slope of a line perpendicular to that line is $\frac{2}{1}$. Just call it 2.

Now that you have the slope of the new line, you need to find the y-intercept for the new line. Use your old friend $y = mx + b$.

Substitute in what you know:

$$4 = 2(5) + b$$

$$4 = 10 + b$$

$$b = -6$$

The line that is perpendicular to the original line has a y-intercept of –6.

Now write the equation for that line:

$$y = mx + b$$

$$y = 2x - 6$$

This line is perpendicular to $y = -\frac{1}{2}x + 3$. Check out this graph of the two lines.

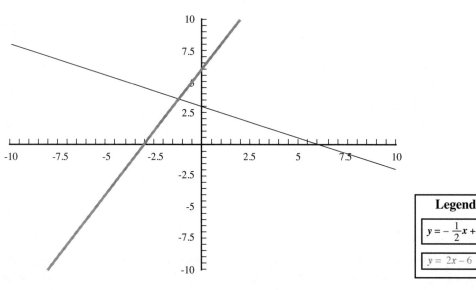

Legend
$y = -\frac{1}{2}x + 3$
$y = 2x - 6$

Graph of $y = -\frac{1}{2}x + 3$ and $y = 2x - 6$.

Now It's Your Turn

Use your knowledge of parallel and perpendicular lines to answer each of the following questions. You can find the answers in Appendix A.

21. Are these two lines parallel? Explain your answer.

    $3y = 2x - 9$

    $12y = 8x + 19$

22. Are these two lines perpendicular? Explain your answer.

    $y = 10x - 1$

    $-10y - 2x = 8$

23. What is the equation for a line that passes through (5,7) and is parallel to the line $y = \frac{2}{5}x - 1$?

24. What is the equation for a line that passes through (5,7) and is perpendicular to the line $y = \frac{2}{5}x - 1$?

25. A line passes through (8,6) and (6,–2). Is this line perpendicular to a line that passes through (–6,6) and (2,6)? Explain.

You've Got the Data, Where Is the Line?

Now you know a lot of different ways to describe, graph, and write linear equations. Let's put all that hard work to practical use.

Scatter Plots and Their Lines

Scatter plots are used to show a relationship between pairs of data. In many cases, it is important to make a line of best fit of these data points to show a trend in the data you have gathered. This is a line that is drawn which connects an "average" of all the points. Not all the points will be on the line and that is ok. The idea is to draw a line to represent the points as best as possible—a line of best fit. Look at the example in the following section.

Example 9

A high school track coach wants to track how her team's times have improved during the course of 6 weeks. She makes the following table.

Week	Time (Seconds)
1	63
2	61
3	62
4	60
5	59
6	57

In this case, you can plot the data on a graph. Use the values of week for the x-axis and the times on the y-axis.

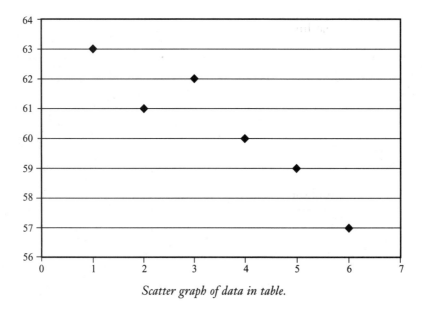

Scatter graph of data in table.

This is a scatter graph. Look at the points on this graph. See how you can draw a line that connects them? The line might not touch every point, but the line represents an overall trend. Draw that line.

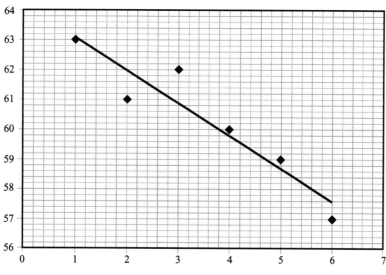

Scatter graph of data in table with best fit line.

Now, choose two points on that line and find the slope of the line using the formula $m = \dfrac{y_2 - y_1}{x_2 - x_1}$.

Now, write the equation:

$y = mx + b$

Try another example.

Example 10

Use the data in this table to find the equation of the best fit line.

x	y
1	3
2	5
3	4
4	7
5	10
6	11
7	12

Graph the scatter plot of this data.

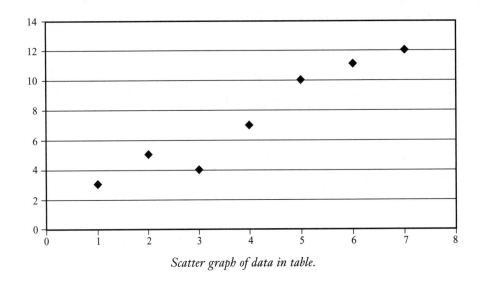

Scatter graph of data in table.

Draw a line that fits this data.

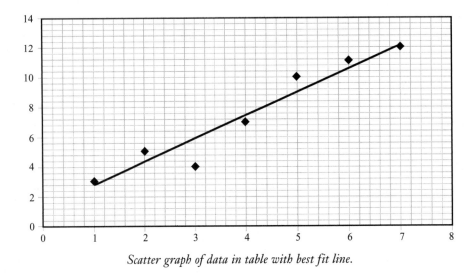

Scatter graph of data in table with best fit line.

Choose two points on the line and determine the slope. Write the equation for the line.

Now It's Your Turn

Write the equation which corresponds with each set of data. The solutions you derive will be approximations, which should come close to the exact values found in Appendix A.

26. A football team of second and third graders plays eight games in a season. The number of points they scored in each game is in the table that follows.

Week	Points Scored
1	12
2	0
3	13
4	14
5	6
6	21
7	0
8	7

Write the equation of the line that best fits this data.

27. A family keeps track of the number of different bird species that arrive at their bird feeder each month. The results are shown in the table that follows.

Month	Number
September	12
October	8
November	15
December	16
January	18
February	20

Write the equation of the line that best fits this data.

28. Make a scatter plot for the following data and write the equation for the best fit line.

x	y
1	3
2	4
3	8
4	7
5	5
6	9
7	5

29. Make a scatter plot for the following data and write the equation for the best fit line.

x	y
1	10
2	7
3	5
4	−3
5	−4
6	−8
7	−10

30. Make a scatter plot for the following data and write the equation for the best fit line.

x	y
0.5	15
1	12
1.5	34
2	49
2.5	53
3	76
3.5	112

Chapter

8

Moving On to Linear Inequalities

In This Chapter

- Solving inequalities
- Graphing inequalities
- Understanding special inequalities and their graphs
- Graphing inequalities with two variables

Now that you are an expert in linear equalities, it is time to switch things up. This chapter will look at linear inequalities. Keep in mind that these are similar to equalities, except they don't contain that comforting equal sign. Inequalities can get a little competitive with each other. They are all about who is greater and who is less than the other. Let's take a look.

Inequalities

These are those sentences in algebra that have <, >, ≤, and ≥ symbols rather than an equal sign.

> **LOOK OUT!**
>
> Remember that an equality is a mathematical statement that has an = between two expressions. Inequalities have symbols such as <, >, ≤, and ≥ between two expressions.

Don't be worried. Inequalities are not difficult. Many of the same rules apply that apply to plain old equations. The only thing different is that you have to keep track of those symbols. With a little practice, that will all work out fine.

Adding and Subtracting Inequalities

When adding or subtracting inequalities, it is helpful to keep the addition property of inequality and the subtraction property of inequality in mind. The addition property of inequality says that if you add the same number to each side of an inequality, the result is an equivalent inequality.

Think of it this way:

If $a > b$, then $a + c > b + c$.

This is true for any of the inequalities.

The subtraction property of inequality says that if the same number is subtracted from each side, the result will be an equivalent inequality.

In other words, if $x < y$, then $x - z < y - z$.

Maybe working through a couple of examples will help make you more comfortable.

Example 1

Solve $x + 4 > 8$.

What do we have here? This inequality says that any number added to 4 is greater than 8. So the goal here is to learn more about the value of x. To do this, subtract 4 from each side (using the subtraction property of inequality).

$$x + 4 > 8$$
$$x + 4 - 4 > 8 - 4$$
$$x > 4$$

So any number greater than 4 makes this correct. You can check it by plugging in a few numbers for x into the original inequality.

Try plugging in 2 (which is less than 4):

$$2 + 4 > 8 \text{ or } 6 > 8. \text{ That's not true.}$$

Try plugging in 5 (which is greater than 4):

$$5 + 4 > 8 \text{ or } 9 > 8. \text{ That's right!}$$

Example 2

Try another one.

$$y - 16 \leq 4$$

To get y by itself, you need to add 16 to each side of the inequality:

$$y - 16 + 16 \leq 4 + 16$$
$$y \leq 20$$

Multiplying and Dividing Inequalities

Multiplying and dividing is not that difficult but you have to watch for some slightly tricky things:

- If you multiply both sides of an inequality by a positive number, the inequality symbol stays the same as it was before.

- If you multiply each side on an inequality by a negative number, you have to switch the direction of the inequality symbol to produce an inequality that is equivalent.

A similar thing happens for dividing inequalities:

- Dividing both sides by a positive number creates an equivalent inequality without anything having to change.

- Dividing by a negative number requires that the inequality symbol is switched.

> **PSST—TRY IT THIS WAY**
>
> If you are a visual learner, you might find these formulas useful.
>
> When multiplying inequalities:
>
> $$\text{If } a < b \text{ and } c > 0, \text{ then } ac < bc.$$
> $$\text{If } a < b \text{ and } c < 0, \text{ then } ac > bc.$$
>
> When dividing inequalities:
>
> $$\text{If } a < b \text{ and } c > 0, \text{ then } \frac{a}{c} < \frac{b}{c}.$$
> $$\text{If } a < b \text{ and } c < 0, \text{ then } \frac{a}{c} > \frac{b}{c}.$$
>
> This applies for all the inequality symbols.

Example 3

Solve this inequality using division:

$$-5x < 35$$

This inequality says that some number, when multiplied by –5, is less than 35.

So what could those numbers be? Divide each side of the equation by –5 to get x by itself:

$$\frac{-5x}{-5} < \frac{35}{-5}$$
$$x > -7$$

Did you catch that? You have to remember to reverse the inequality symbol because you are dividing by a negative number. Let's check if this is right.

Any number greater than –7 should make the inequality correct. If –2 is plugged into the original equation, is that inequality correct? That's …

$$(-5)(-2) < 35$$

$$10 < 35. \text{ Yes, that is right!}$$

Multi-Step Inequalities

Adding and subtracting or multiplying and dividing inequalities is not so bad. Those are single-step problems where one number has to be moved to the other side of the equation and then the problem is solved. But things can't always be that neat. So next up is a look at how to solve inequalities that involve multiple steps.

Example 4

Solve the following:

$$4x - 6 \geq 22$$

In plain English, this inequality says that any number multiplied by 4 and then from which 6 is subtracted is greater than or equal to 22. That's quite a mouthful.

The first step is to work to get the variable by itself. The logical first step is to add 6 to each side. Leave the inequality symbol alone because you are just adding here:

$$4x \geq 22 + 6$$

$$4x \geq 28$$

Now divide each side by 4 (positive 4 that is!):

$$x \geq 7$$

Okay, that was only a two-step inequality. But there can be even more steps than that.

Example 5

Solve this multi-step inequality:

$$9x - 7 > 3x + 17$$

Gulp. Okay, one step at a time. Add 7 to each side of the equation:

$$9x > 3x + 17 + 7$$

$$9x > 3x + 24$$

Work to get all the x variables on the same side. So subtract $3x$ from each side:

$$9x - 3x > 3x - 3x + 24$$

$$6x > 24$$

$$x > 4$$

Whew. That wasn't so bad. Just take it one step at a time!

Now It's Your Turn

Solve the following inequalities. The answers are located in Appendix A.

1. $3y + 9 \leq 12$

5. $-3(6y + 5) \geq -22$

2. $4x - 2 > 7$

6. $4y - 5 > 3y + y - 7$

3. $18 - 6y \geq 28$

7. $3(b + 6) < 48$

4. $4p - 3 \leq 7p + 7$

8. $5(t - 6) > 4t - 16$

Graphing Inequalities

You've solved inequalities but you haven't seen them yet. What do graphs of inequalities look like? Let's see.

PSST—TRY IT THIS WAY

Inequalities, as shown here, are graphed on a number line. There is no need for a large *x-y* graph as has been used before.

Example 6

Graph the solution to the following inequality on a number line:

$2x + 6 > 16$

First, solve the inequality. Subtract 6 from each side:

$$2x > 16 - 6$$

$$2x > 10$$

Now solve for *x*:

$$x > 5$$

So the solution to the inequality is any number that is greater than 5 (not including 5). On a number line, there is an open circle around 5 and an arrow pointing to the right (to indicate numbers greater than 5).

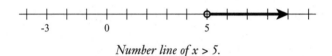

Number line of x > 5.

What if an inequality has a greater than and equal to symbol (≥) or the less than and equal to symbol (≤)? How is that graphed?

Example 7

Graph the solution to the following inequality on a number line:

$$10 \leq x + 6$$

Solve for *x* by subtracting 6 from both sides:

$$10 - 6 \leq x + 6 - 6$$

$$4 \leq x$$

So any number that is greater than or equal to 4 will solve the inequality when graphed; this solution looks like this:

Number line of 4 ≤ x.

Now It's Your Turn

Graph the solution to each of the inequalities you solved in the last section on a number line. You can locate the answers in Appendix A.

9. $3y + 9 \leq 12$

13. $-3(6y + 5) \geq -22$

10. $4x - 2 > 7$

14. $4y - 5 > 3y$

11. $18 - 6y \geq 28$

15. $3(b + 6) < 48$

12. $4p - 3 \leq 7p + 7$

16. $5(t - 6) > 4t - 16$

Special Inequalities

As you might have guessed, now that you are feeling comfortable with inequalities, not every possible case is this easy. There are several "special cases" that need to be addressed before moving on.

Compound Inequalities

A compound inequality has two different inequalities that are related by the term *and* or the term *or*. For example, you are in a coffee shop with your favorite latte and you overhear someone say, "What if I had all real numbers that are greater than 7 and less than 12?" Being an overzealous eavesdropper, you would immediately know that an inequality can be written to represent that statement.

Example 8

Write the inequality for the statement you overheard and graph it. What if I had all real numbers that are greater than 7 and less than 12?

$7 < x < 12$

Plotted on a number line, that inequality looks like this:

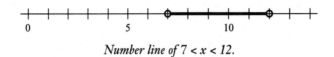

Number line of 7 < x < 12.

That example was not too bad. Take a look at another one!

Example 9

Solve the inequality and then graph the solution:

$8z + 2 \leq -6$ or $10z - 6 > 34$

Deep breath. Solve each inequality separately. That will make things more manageable.

$8z + 2 \leq -6$

Subtract 2 from each side:

$8z + 2 - 2 \leq -6 - 2$

$8z \leq -8$

Solve for z by dividing both sides by 8:

$z \leq -1$ (good)

Now solve the other inequality for z.

$10z - 6 > 34$

Add 6 to each side:

$10z - 6 + 6 > 34 + 6$

That leaves us with:

$10z > 40$

Solve for z:

$z > 4$

The solution for the inequality is all real numbers less than or equal to –1 or less than 4. Graphed on a number line, the solution looks like this:

Number line of z ≤ –1 or z > 4.

Now It's Your Turn

Solve the following inequalities and graph the solution on a number line. The answers can be found in Appendix A.

17. $3 < x + 7 < 9$

19. $4x + 6 < 18$ or $6x – 12 > 24$

18. $-8 \leq y - 5 < 12$

20. $4n + 2 < -9$ or $3n - 7 > 7$

Solving Inequalities with Absolute Values

You can't avoid it. There are going to be inequalities with absolute values as well.

Example 10

Solve the following absolute inequalities:

$|z| \leq 6$

In this example, the distance between z and 0 on the number line is less than or equal to 6. This means that $-6 \leq z \leq 6$.

On a number line it would look like the following:

Number line of $-6 \leq z \leq 6$.

Here's another one for you.

$$|k| \geq 0.25$$

In this example, the distance between k and 0 is greater than or equal to 0.25. So, $k \leq -0.25$ or $k \geq 0.25$.

Graphed on a number line, the solution looks like this:

Number line of $k \leq -0.25$ or $k \geq 0.25$.

Now It's Your Turn

Solve the following inequalities. Use the same rules that you used for other inequalities. Graph the solutions on a number line. The answers are located in Appendix A.

21. $|y + 5| < 8$

22. $4|2z - 3| - 5 < 11$

23. $|r + 8| \geq 19$

24. $|5b - 7| \leq 20$

25. $|b| \geq -9$

26. $4 + m \geq 0$

Graphing Inequalities with Two Variables

This chapter is almost over. It's time to really mix things up. When graphing an inequality with two variables, you are going to pretend. Take the inequality and change it to an equation by changing the inequality symbol to an equal sign. Then … oh, let's work through the example and see what happens.

Example 11

Graph the inequality $y > 8x - 10$

The first thing you do is to change that inequality into an equation. The equation is now

   $y = 8x - 10$

Graph that line.

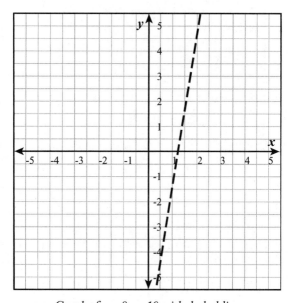

Graph of y = 8x − 10 with dashed line.

LOOK OUT!

Rules to remember:

When graphing an inequality with a > or < symbol, use a dashed line.

When graphing an inequality with a ≥ or ≤ symbol, use a solid line.

Test the point (0,0) in the inequality. When substituted in, the inequality is now:

   $0 > -10$ and that is correct!

So shade in the side of the dashed line that contains (0,0).

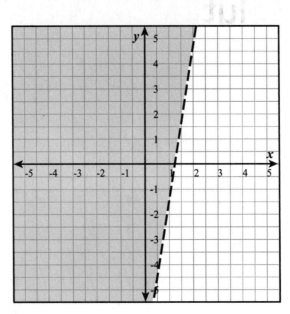

Graph of y = 8x − 10 with dashed line and shading.

Now It's Your Turn

Graph the following inequalities. Don't forget the shading! You can find the answers in Appendix A.

27. $y < x - 2$

29. $y < 4x + 7$

28. $y \geq -3x + 7$

30. $4x - y \geq -1$

Solution Sets and Matrix Math

Time for a little deep breathing. You are now ready to solve not just one equation or inequality—oh no, you are poised on the brink of solving sets of them. Several at one time. After that you will dive into solving different sets of numbers—those found in the matrix (which unfortunately does not involve a special appearance by Keanu Reeves).

Linear Equations, Inequalities, and Bears ... Oh My!

In This Chapter

- Solving solution sets
- Dealing with solution sets with no solution or too many solutions
- Solving systems of two linear equations
- Solving systems of three linear equations

How to Solve Linear Systems—by Graphing

Give you a linear equation and you'd be more than happy to graph it, wouldn't you? You mastered that in Chapter 5. But what if suddenly you were asked to graph to solve a linear system? What on Earth does that involve? Well, I am glad you asked.

A linear system has two more linear equations with the same variables. If you solve the linear system, the answer will be an ordered pair that satisfies each equation. It's probably best to start with an example.

Example 1

Here is a linear system:

$$x + 3y = 8$$
$$2x - 4y = 8$$

Notice how each has the same variables? Graph each of those equations on the same graph, as follows:

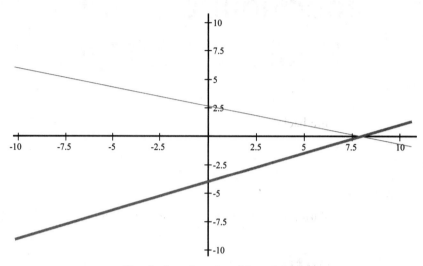

Graph of x + 3y = 8 and 2x − 4y = 8.

The point where the lines intersect is a solution for this linear set. As you can see, the lines intersect at (5.6,.8), which is a solution. Check it by substituting in the value for *x* or *y* into both of the equations.

Solving by Substitution

Graphing is not the only way to solve a linear system. They can be solved by substitution as well. This way is pretty fun.

Example 2

Solve the linear system using substitution:

$y = 2x + 4$

$x + 3y = 26$

Arbitrarily call the first equation A and the second equation B. It makes life easier to assign names. Solve one of the equations for *y*. And hey! Look at that, Equation A is already solved for *y*. Now substitute Equation A into Equation B for the value of *y*. Like this:

$x + 3(2x + 4) = 26$

Now solve for x:

$$x + 6x + 12 = 26$$
$$7x = 14$$
$$x = 2$$

Use this value of x to substitute back into Equation A to get the value of y:

$$y = 2(2) + 4$$
$$y = 8$$

And, drum roll please, the solution to this set of linear equations is (2,8).

Solving by Elimination

The possibilities are seemingly endless when it comes to solving linear systems. The other possible way is by elimination. Use this method if you can find a way to add or subtract the equations to get down to only one variable.

Let's take a look.

Example 3

Solve the following linear system using elimination:

$$7x + 4y = 17$$
$$-7x + y = 13$$

Notice the x variables? One has the coefficient 7 and the other -7? If you were to add these two equations together, you'd eliminate the x variable.

Then you'd be left with $5y = 30$. Solving for y the answer is $y = 6$.

Substitute y into either of the equations to get the value of x:

$$7x + 4(6) = 17$$
$$7x + 24 = 17$$
$$7x = -7$$
$$x = -1$$

The solution set for this pair is $(-1,6)$.

Now It's Your Turn

Solve the following linear system by graphing. The answers are available in Appendix A.

1. $2x - y = -11$

    $y = -2y - 13$

2. $x - 2 = 3$

    $2x + 3y = 22$

3. $x + 2y = 10$

    $-x + y = 5$

Solve the following systems by substitution.

4. $x - 2y = 12$

    $4x + y = -1$

5. $y = 4x - 6$

    $2x + 5y = 12$

6. $6x + 2y = 39$

    $4x + 2y = 18$

7. $3x + y = 15$

    $y = -15$

Solve the following systems by elimination.

8. $4x - 8y = -18$

    $4x + 6y = -9$

9. $-7x + 6y = 28$

    $6x - y = 12$

10. $4x + 9y = 33$

    $-4x + 7y = 2$

Special Cases

As always, there are going to be certain linear systems that do not fit the mold. Some linear systems have no answers. And some have too many!

Systems with No Answers

You'd think this would be a good thing. Hey, if there is no answer, there's simply no answer. Unfortunately, most teachers (and algebra book authors) are going to be sticklers and want to know *why* there is no answer. So, be ready ...

Example 4

Show that this linear system has no answer:

$$6x + 4y = 20$$

$$6x + 4y = 4$$

You could prove this in any number of ways. Let's see why that is the case by graphing. If these two equations are graphed this is what you get:

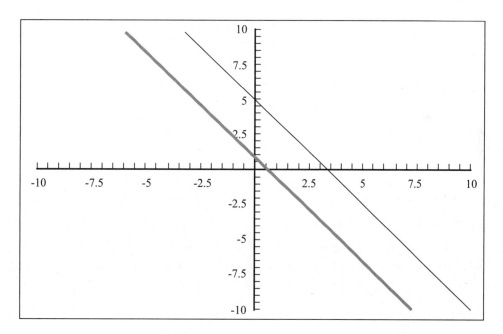

Graph of 6x + 4y = 20 and 6x + 4y = 4.

The lines are parallel. They have the same slope! This means that they never meet. Pretty difficult to find the point where the lines intersect, if they never meet. This proves that there is no answer to that linear system.

You could also do it by elimination. Let's try to solve it that way, too (just for fun):

$$6x + 4y = 20$$
$$6x + 4y = 4$$

If you subtract those two equations you would get …

$$0 = 16$$

In most places, 0 does not equal 16. There is no solution to this linear system.

Linear Systems with Too Many Answers

This was often my problem in algebra, years ago. There were too many answers. What does this mean? Whatever the answer, this is certainly a special case in algebra.

Example 5

Show that this linear system has too many solutions (an infinite amount actually):

$$y = -5x - 2$$
$$-20x - 4y = 8$$

Solve this by substitution. Replace y in the second equation with $y = -5x - 2$.

Now it reads …

$$-20x - 4(-5x - 2) = 8$$
$$-20x + 20x + 8 = 8$$
$$8 = 8$$

So, no matter what value is substituted in here, the value of y is always going to be the same. These equations have the same slope and the same y-intercept. There is an infinite number of solutions here.

On a graph, this is what these two equations look like.

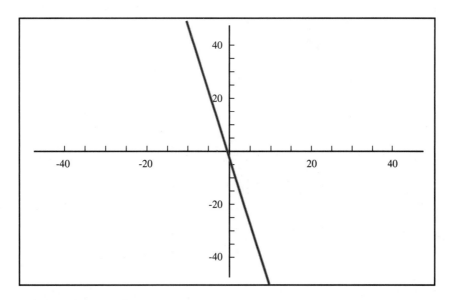

Legend

$-20x - 4y = 8$

$y = -5x - 2$

Graph of $-20x - 4y = 8$ and $y = -5x - 2$.

PSST—TRY IT THIS WAY

Here is a helpful chart that might make determining the number of solutions to a linear system easier!

Things to Look For ...	Number of Solutions
Different slopes	One correct solution
Same slope, different y-intercept	No solution
Same slope, same y-intercept	An infinite number of solutions

Now It's Your Turn

Solve each linear system and indicate if there is one solution, no solutions, or an infinite number of solutions. You can find the answers in Appendix A.

11. $6x - 8y = 24$

$y = \dfrac{3}{4}x - 3$

12. $-2x + 2y = -12$

$6x - 12y = -30$

13. $-x + y = 12$

    $x - y = 12$

14. $-x + y = 7$

    $x - y = -7$

15. $15x - 5y = 25$

    $6x - 2 = 8$

16. $y = 7x + 13$

    $-7x + 3y = 39$

17. $2x - 2y = -8$

    $-6x + 6y = 8$

18. $3x - 3y = -15$

    $4x - 4y = -4$

19. $6x + y = 24$

    $5x - y = 12$

20. $x + 4y = -1$

    $-2x - 3y = 2$

Systems of Inequalities

Solving systems of inequalities is similar to solving systems of equalities. The symbols are different, that is all. The graph results in a darker shaded region as you can see in the examples.

Example 6

Graph this system of two inequalities (and hang on, there can be more than two inequalities!):

$$y > -\frac{1}{3}x - 6$$

$$y \leq \frac{1}{2}x + 2$$

Remember when you are going to graph these things (inequalities) to start out by graphing the equation for it as if it were an equality. Test a point, and then shade in the area in which that point lies. The line is a dotted line if the inequality is > or <. The line is solid if the inequality contains a ≥ or ≤ symbol.

Here is the graph of these two inequalities.

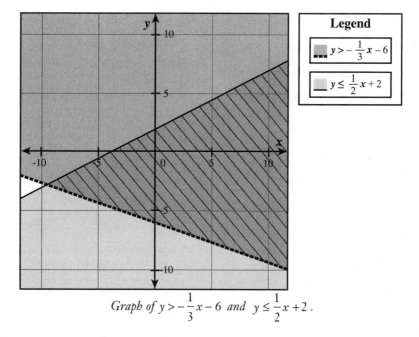

Legend

$y > -\frac{1}{3}x - 6$

$y \le \frac{1}{2}x + 2$

Graph of $y > -\frac{1}{3}x - 6$ and $y \le \frac{1}{2}x + 2$.

Notice how there is some overlap in the shaded regions. It is the overlapped area that is your actual answer for the graph of this system of inequalities. Try another one.

Example 7

Graph this system of inequalities:

$y < 3x + 6$

$y \ge -x + 4$

Graph each inequality separately on the same graph, shading in the portion that each inequality represents.

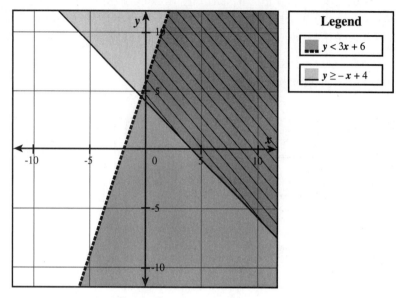

Graph of y < 3x + 6 and y ≥ –x + 4.

The shaded overlapped portion is the answer!

Sometimes, algebra teachers and algebra book authors get a little carried away with systems of linear inequalities and they might throw a problem at you that has three linear inequalities to solve. There is no need to panic or run away. Let's walk through a few examples now.

Example 8

Graph this system of inequalities:

$$y > -3$$

$$x \geq -1$$

$$x + 2y < 5$$

Yikes. Take the inequalities one at a time. Start with $y > -3$. When graphed, it looks like this:

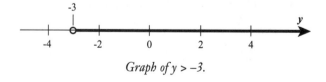

Graph of y > –3.

No problem. Now add the graph $x \geq -1$. What, no y in this inequality? It must be a vertical line!

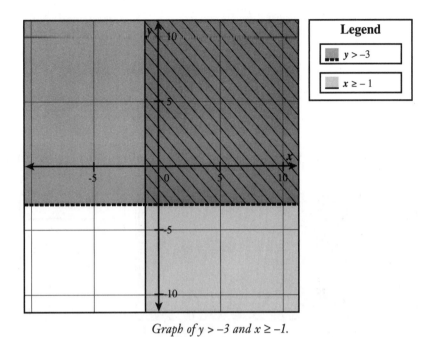

Graph of y > –3 and x ≥ –1.

Okay, so there is some overlap here. Good. Add the third inequality to these two. Be sure to rearrange to solve for *y* with the inequality $x + 2y < 5$.

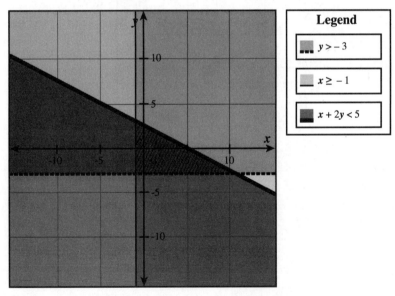

Graph of y > –3 and x ≥ –1 and x + 2y < 5.

And there you have it, the solution region of the three inequalities! Try another one. You know you love the challenge.

Example 9

Graph this system of inequalities:

$$y \geq -x + 8$$

$$y > 3$$

$$x < -5$$

Now, do the same thing with one inequality at a time. The final result looks like this:

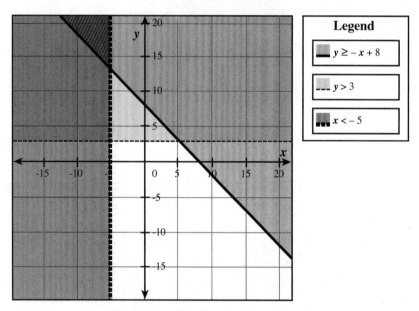

Graph of y ≥ –x + 8 and y > 3 and x < –5.

Example 10

One last one. Graph this system of inequalities:

$$x > 8$$

$$y > x$$

$$y < 2x - 7$$

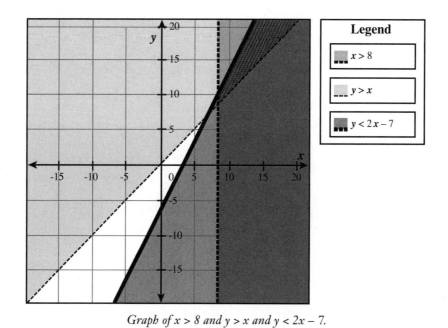

Graph of x > 8 and y > x and y < 2x – 7.

Now It's Your Turn

Graph the following systems of inequalities. You can find the answers in Appendix A.

21. $x > 5$

    $x < -2$

24. $x + 6y > 12$

    $x \geq 4$

22. $y \leq 12$

    $y \geq 8$

25. $x + y \leq 14$

    $x - y < 7$

    $y > 3$

23. $y \leq 2x + 4$

    $y > -x + 5$

26. $x \leq 0$

    $y < 0$

    $5y - x > 15$

27. $x > 6$

    $y \leq 2$

    $y \geq -3x + 1$

29. $y < 3x + 8$

    $y \geq x$

    $y \geq 4x - 8$

28. $x + y \geq 2$

    $2x + y < 9$

    $y \leq 1$

30. $y < -2$

    $2x + y > 10$

Welcome to the Matrix

Every so often, a strange sight appears in algebra books. If you see something that looks like this, you are looking at a matrix.

$$\begin{bmatrix} 12 & -16 & 10 \\ 0 & 47 & 0 \\ -9 & -1 & 2 \end{bmatrix}$$

In this chapter, you will learn to add, subtract, and multiply matrices. It can be a little complicated, but if you follow the steps, it won't be that hard.

The Matrix

A matrix is an arrangement of numbers in rows and columns. The rows are usually labeled m and the columns as n. This means that the dimension of the matrix is equal to the number of rows multiplied by the number of columns. It is important to know the dimensions of a matrix to perform certain operations on the matrix. Let's walk through an example to get you oriented in matrix space.

Example 1

Examine the matrix that follows and identify the *dimension* and the *elements*.

$$\begin{bmatrix} 4 & -1 & 6 \\ 7 & 0 & 2 \end{bmatrix}$$

In this matrix, the dimension is 2×3. The element in the first row and third column is 6. The element in the second row and second column is 0.

DEFINITION

The **elements** of a matrix are the numbers. Each matrix is defined by its **dimension** or rows and columns. Any matrix has m rows and n columns. The previous matrix has a dimension of 2×3.

Scalpel please. We are about ready to start performing basic operations on matrices.

Adding Matrices

A few things about adding two matrices. You can only add matrices with the same dimensions. If the dimensions are the same, then you add the elements in the corresponding position. Let's try an example.

PSST—TRY IT THIS WAY

It may help to remember the addition of two matrices as follows:

$$\begin{bmatrix} w & x \\ y & z \end{bmatrix} + \begin{bmatrix} a & b \\ c & d \end{bmatrix} = \begin{bmatrix} w+a & x+b \\ y+c & z+d \end{bmatrix}$$

Example 2

Add these two matrices:

$$\begin{bmatrix} 2 & 5 \\ -1 & 0 \end{bmatrix} + \begin{bmatrix} 8 & 9 \\ 4 & -6 \end{bmatrix}$$

Add the elements in the corresponding positions:

$$\begin{bmatrix} 2+8 & 5+9 \\ -1+4 & 0+-6 \end{bmatrix}$$

This is equal to:

$$\begin{bmatrix} 10 & 14 \\ 3 & -6 \end{bmatrix}$$

Example 3

Try another one:

$$\begin{bmatrix} 1 & 8 & -7 \\ 8 & 4 & 13 \\ 0 & -1 & 2 \end{bmatrix} + \begin{bmatrix} 0 & 3 & 15 \\ 1 & -5 & 5 \\ 3 & 6 & -2 \end{bmatrix}$$

Same thing here. Add the elements in the corresponding positions:

$$\begin{bmatrix} 1+0 & 8+3 & -7+15 \\ 8+1 & 4+-5 & 13+5 \\ 0+3 & -1+6 & 2+-2 \end{bmatrix} = \begin{bmatrix} 1 & 11 & 8 \\ 9 & -1 & 18 \\ 3 & 5 & 0 \end{bmatrix}$$

Subtracting Matrices

Subtracting matrices is not too bad either.

Again, you need to ensure that the dimensions of the two matrices are the same. And if they are, then subtract the elements in the corresponding positions.

PSST—TRY IT THIS WAY

It may help to remember the subtraction of two matrices as follows:

$$\begin{bmatrix} w & x \\ y & z \end{bmatrix} - \begin{bmatrix} a & b \\ c & d \end{bmatrix} = \begin{bmatrix} w-a & x-b \\ y-c & z-d \end{bmatrix}$$

Example 4

Subtract these two matrices:

$$\begin{bmatrix} 5 & 8 & 10 \\ -6 & 2 & 8 \\ 1 & 0 & -7 \end{bmatrix} - \begin{bmatrix} 4 & 9 & 7 \\ -7 & 5 & 0 \\ 2 & -11 & 8 \end{bmatrix}$$

Subtract the elements in the corresponding positions:

$$\begin{bmatrix} 5-4 & 8-9 & 10-7 \\ -6-(-7) & 2-5 & 8-0 \\ 1-2 & 0-(-11) & -7-8 \end{bmatrix}$$

This is equal to …

$$\begin{bmatrix} 1 & -1 & 3 \\ 1 & -3 & 8 \\ -1 & 11 & -15 \end{bmatrix}$$

LOOK OUT!

Remember that you cannot add and subtract a set of matrices unless the dimension is the same. Always check that first before working out a problem!

Now It's Your Turn

Try your hand at a few now. Add or subtract each matrix, if possible. The answers are located in Appendix A.

1. $\begin{bmatrix} 9 & 8 \\ 2 & 4 \end{bmatrix} + \begin{bmatrix} -7 & 0 \\ 1 & -10 \end{bmatrix}$

5. $\begin{bmatrix} 4 & -5 & 1 \\ 3 & 9 & -17 \\ 11 & 0 & -3 \end{bmatrix} + \begin{bmatrix} -6 & 0 & 12 \\ -2 & 10 & 7 \\ -1 & 3 & 12 \end{bmatrix}$

2. $\begin{bmatrix} -3 & 12 \\ 9 & -4 \end{bmatrix} - \begin{bmatrix} 5 & -2 \\ 11 & -8 \end{bmatrix}$

6. $\begin{bmatrix} 3 & 12 & -8 & -4 \\ 0 & 1 & 11 & 9 \end{bmatrix} + \begin{bmatrix} 3 & 12 & 15 & -19 \\ 10 & 20 & -16 & 3 \end{bmatrix}$

3. $\begin{bmatrix} 2 & 8 & 17 \\ -5 & -9 & 1 \\ 0 & 3 & 4 \end{bmatrix} + \begin{bmatrix} 6 & 8 \\ 4 & 2 \end{bmatrix}$

7. $\begin{bmatrix} 2 & 5 \\ 1 & 3 \\ 7 & 2 \end{bmatrix} - \begin{bmatrix} -9 & -8 \\ 10 & 12 \\ -7 & 0 \end{bmatrix}$

4. $\begin{bmatrix} 9 & 4 & 1 \\ -6 & 0 & -8 \\ 1 & 32 & 7 \end{bmatrix} - \begin{bmatrix} 1 & 0 & 2 \\ -1 & 12 & 3 \\ -8 & 14 & 2 \end{bmatrix}$

8. $\begin{bmatrix} 2 & 12 & 9 & 7 \\ 10 & -8 & -1 & -22 \\ 3 & 5 & -10 & 4 \\ 0 & 3 & -17 & -6 \end{bmatrix} + \begin{bmatrix} 3 & 1 & 0 & -2 \\ 9 & 4 & 13 & -4 \\ 6 & -8 & -1 & 2 \\ 8 & 5 & 12 & 0 \end{bmatrix}$

Multiplying Matrices

Adding and subtracting matrices might have been a little labor-intensive and you had a lot of numbers to keep track of, but now you have to multiply them.

There are two different types of multiplication that can occur with a matrix. The first one is when a matrix is multiplied by a *scalar*. Let's take a look at the next example.

> **DEFINITION**
>
> A **scalar** is a real number. In this matrix, the scalar is 2.
>
> $$2\begin{bmatrix} 2 & 7 & 1 \\ 0 & 6 & 11 \\ -4 & 9 & 3 \end{bmatrix}$$

Example 5

Multiply the matrix:

$$4\begin{bmatrix} 1 & 8 & -1 \\ 2 & 6 & 11 \\ -4 & 10 & -3 \end{bmatrix}$$

Each element in the matrix is multiplied by the scalar (4 in this example):

$$4\begin{bmatrix} 1 & 8 & -1 \\ 2 & 6 & 11 \\ -4 & 10 & -3 \end{bmatrix} = \begin{bmatrix} 4(1) & 4(8) & 4(-1) \\ 4(2) & 4(6) & 4(11) \\ 4(-4) & 4(10) & 4(-3) \end{bmatrix}$$

This is equal to:

$$\begin{bmatrix} 4 & 32 & -4 \\ 8 & 24 & 44 \\ -16 & 40 & -12 \end{bmatrix}$$

Of course, not all multiplication is that straightforward. There are certain rules that must be followed when multiplying two matrices. First and foremost, the number of columns in the first matrix must be equal to the number of rows in the second matrix.

When multiplying two matrices, you multiply each element in the column of the first matrix by the corresponding element in the column of the second matrix. Sometimes it's easier to learn by doing.

Example 6

Find the product of these two matrices:

$$\begin{bmatrix} 2 & 1 \\ 7 & 4 \end{bmatrix}\begin{bmatrix} -5 & 3 \\ 6 & 8 \end{bmatrix}$$

Call $\begin{bmatrix} 2 & 1 \\ 7 & 4 \end{bmatrix}$ Matrix A and $\begin{bmatrix} -5 & 3 \\ 6 & 8 \end{bmatrix}$ Matrix B.

To multiply, you multiply the elements in the first row of Matrix A with the elements in the first column of Matrix B. In other words, the 2 in Matrix A is multiplied by –5 in Matrix B. The 1 in Matrix A is multiplied by 6 in Matrix B. These numbers are added together, and the result is put in the first column, first row of the answer.

$$\begin{bmatrix} 2 & 1 \\ 7 & 4 \end{bmatrix}\begin{bmatrix} -5 & 3 \\ 6 & 8 \end{bmatrix} = \begin{bmatrix} 2(-5)+1(6) & \\ & \end{bmatrix}$$

But you are not done. Now multiply the numbers in the first row of Matrix A with the numbers in the second column of Matrix B. Add the numbers and place that in the first row, second column of the answer. So that it looks like this:

$$\begin{bmatrix} 2 & 1 \\ 7 & 4 \end{bmatrix}\begin{bmatrix} -5 & 3 \\ 6 & 8 \end{bmatrix} = \begin{bmatrix} 2(-5)+1(6) & 2(3)+1(8) \\ & \end{bmatrix}$$

You aren't done yet. Next multiply the numbers in the second row of Matrix A with the numbers in the first column of Matrix B, add them, and put the result in the second row, first column of the answer.

One important rule when it comes to multiplying matrices is this. The number of columns in the first matrix must be equal to the number of rows in the second one.

$$\begin{bmatrix} 2 & 1 \\ 7 & 4 \end{bmatrix}\begin{bmatrix} -5 & 3 \\ 6 & 8 \end{bmatrix} = \begin{bmatrix} 2(-5)+1(6) & 2(3)+1(8) \\ 7(-5)+4(6) & \end{bmatrix}$$

You guessed it, now multiply the second row in Matrix A by the second column in Matrix B, add them together, and place that answer in the second row, second column of the answer.

$$\begin{bmatrix} 2 & 1 \\ 7 & 4 \end{bmatrix}\begin{bmatrix} -5 & 3 \\ 6 & 8 \end{bmatrix} = \begin{bmatrix} 2(-5)+1(6) & 2(3)+1(8) \\ 7(-5)+4(6) & 7(3)+4(8) \end{bmatrix}$$

Now simplify to get this:

$$\begin{bmatrix} -4 & 14 \\ -11 & 53 \end{bmatrix}$$

Whew. Ready to try one more before moving along on your own?

Example 7

Find AB if $A = \begin{bmatrix} 2 & -1 \\ -3 & 2 \end{bmatrix}$ and B equals $\begin{bmatrix} 3 & 0 & 1 \\ -3 & 5 & -2 \end{bmatrix}$.

Although the dimensions of these matrices are different, you can still multiply them. That is because the number of columns in Matrix A is equal to the number of rows in Matrix B.

Begin with the first row in Matrix A. Multiply those numbers by the numbers in the first column in Matrix B, just as you did in the previous example. Add the numbers and place the values in the first column, first row of AB:

$$\begin{bmatrix} 2 & -1 \\ -3 & 2 \end{bmatrix}\begin{bmatrix} 3 & 0 & 1 \\ -3 & 5 & -2 \end{bmatrix} = \begin{bmatrix} 2(3)+(-1)(-3) & & \end{bmatrix}$$

Now multiply the first row in A by the middle column in B. Add these numbers and place the values in the first row, second column of AB:

$$\begin{bmatrix} 2(3)+(-1)(-3) & 2(0)+(-1)5 & \end{bmatrix}$$

Next, multiply the first row in A but the last column in B. Add the numbers and place the values in the first row, third column of AB:

$$\begin{bmatrix} 2(3)+(-1)(-3) & 2(0)+(-1)5 & 2(1)+(-1)(-2) \end{bmatrix}$$

Okay, deep breath. Let's go to the second row of A. Multiply that first by the first column in B, then by the second column in B, and lastly by the third column in B. Each time add the numbers and place in the proper location in AB.

Now your matrix AB looks like this:

$$\begin{bmatrix} 2(3)+(-1)(-3) & 2(0)+(-1)5 & 2(1)+(-1)(-2) \\ (-3)3+2(-3) & (-3)0+2(5) & (-3)(1)+2(-2) \end{bmatrix}$$

And finally simplify:

$$\begin{bmatrix} 9 & -5 & 4 \\ -15 & 10 & -7 \end{bmatrix}$$

Now It's Your Turn

Use multiplication to solve the following problems. The answers can be found in Appendix A.

9. $\begin{bmatrix} 2 \\ 3 \end{bmatrix}\begin{bmatrix} 0 & 4 \end{bmatrix}$

11. $\begin{bmatrix} 3 & -1 & 4 \\ 6 & -2 & -3 \\ 4 & 5 & 1 \end{bmatrix}\begin{bmatrix} -1 & 0 & 12 \\ 9 & -4 & 2 \\ 2 & -10 & 7 \end{bmatrix}$

10. $5\begin{bmatrix} 9 & -1 \\ 4 & -7 \end{bmatrix}$

12. $\begin{bmatrix} 6 & -6 & 3 \\ 0 & 1 & -4 \end{bmatrix}\begin{bmatrix} 0 & -5 \\ 10 & 3 \\ -7 & -1 \end{bmatrix}$

Use the following information to solve questions 13–15.

$$A = \begin{bmatrix} 9 & 0 \\ -1 & 1 \\ 4 & -5 \end{bmatrix} \quad B = \begin{bmatrix} 4 & 11 \\ -2 & 5 \end{bmatrix} \quad C = \begin{bmatrix} 3 & -17 \\ 2 & -1 \end{bmatrix}$$

13. Find A(B + C)

14. Find AB + AC

15. Find A (B – C)

Use the following information to solve the questions 16–18.

$$A = 6 \quad B = \begin{bmatrix} 7 & 12 \\ -6 & 0 \\ -4 & -5 \end{bmatrix} \quad C = \begin{bmatrix} -3 & -5 \\ 12 & -9 \end{bmatrix}$$

16. A(B)

17. A (BC)

18. A (C)

Finding Determinants and Using Cramer's Rule

When looking at square matrix (that with the number of rows equal to the number of columns) there is a real number associated with the matrix. This is called the *determinant*.

DEFINITION

The **determinant** of a matrix can be denoted by det A or by $|A|$. The determinant is equal to the difference of the products of the elements on the diagonals.

PSST—TRY IT THIS WAY

Keep this in mind when finding the determinant of a matrix:

$$\det \begin{bmatrix} a & b \\ c & d \end{bmatrix} = \begin{vmatrix} a & b \\ c & d \end{vmatrix} = ad - cb$$

And for a 3 × 3 matrix:

$$\det \begin{bmatrix} a & b & c \\ d & e & f \\ g & h & i \end{bmatrix} = \begin{vmatrix} a & b & c \\ d & e & f \\ g & h & i \end{vmatrix} \begin{matrix} a & b \\ d & e \\ g & h \end{matrix} = (aei + bfg + cdh) - (gec + hfa + idb)$$

Let's take a look at an example. Remember, you are going to find a real number at the end.

Example 8

Find the determinant of this matrix:

$$\det \begin{bmatrix} 3 & -17 \\ 2 & -1 \end{bmatrix} = 3(-1) - 2(-17)$$

$$\det \begin{bmatrix} 3 & -17 \\ 2 & -1 \end{bmatrix} = 31$$

REALITY CHECK

You have been very patient, but by now you are probably wondering what a matrix is used for. There are many potential uses. One might be to find the total cost of something. If three different stores have four different products, all costing different amounts, then it is possible, by using a matrix, to find the total value of the inventory for each store. Any time there are multiple variables to be considered in a number of different examples, a matrix might be useful.

Let's try another example.

Example 9

Find the determinant of this matrix:

$$\det \begin{bmatrix} 3 & 0 & -3 \\ 2 & -6 & 4 \\ 1 & -5 & 7 \end{bmatrix}$$

Remember to repeat the first two columns when multiplying. So it will look like this:

$$\begin{vmatrix} 3 & 0 & -3 \\ 2 & -6 & 4 \\ 1 & -5 & 7 \end{vmatrix} \begin{matrix} 3 & 0 \\ 2 & -6 \\ 1 & -5 \end{matrix}$$

Now multiply:

$$[(3)(-6)(7) + 0(4)(1) + -3(2)(-5)] - [(1)(-6)(-3) + (-5)(4)(3) + 7(2)(0)]$$

Simplify:

$$(-126 + 30 + 0) - [-18 - 60)] = -54$$

Cramer's Rule

It is possible to use determinants to solve a system of linear equations (gulp!). This is done using Cramer's Rule and uses a coefficient matrix.

Let's say you have the following linear system:

$$ax + by = e$$

$$cx + dy = f$$

In this case, the coefficient matrix is $\begin{bmatrix} a & b \\ c & d \end{bmatrix}$.

Nice and simple, right? Now apply that to Cramer's Rule.

Cramer's Rule says that $x = \dfrac{\begin{vmatrix} e & b \\ f & d \end{vmatrix}}{\det A}$ and $y = \dfrac{\begin{vmatrix} a & e \\ c & f \end{vmatrix}}{\det A}$.

Example 10

Use Cramer's Rule to solve this system:

$$8x + 3y = -7$$

$$2x - 2y = -12$$

First thing to do is find the coefficient matrix. For this system that is

$$\begin{bmatrix} 8 & 3 \\ 2 & -2 \end{bmatrix}$$

Now find the determinant of that matrix:

$$\det \begin{bmatrix} 8 & 3 \\ 2 & -2 \end{bmatrix} = \begin{vmatrix} 8 & 3 \\ 2 & -2 \end{vmatrix} = 8(-2) - 2(3)$$

The determinant is equal to $-16 - 6 = -22$.

Now plug all the information you need into Cramer's Rule:

$$x = \frac{\begin{vmatrix} e & b \\ f & d \end{vmatrix}}{\det A} \text{ and } y = \frac{\begin{vmatrix} a & e \\ c & f \end{vmatrix}}{\det A}$$

$$x = \frac{\begin{vmatrix} -7 & 3 \\ -12 & -2 \end{vmatrix}}{-22} = \frac{(-7)(-2) - (-12)3}{-22} = -2.27$$

$$y = \frac{\begin{vmatrix} 8 & -7 \\ 2 & -12 \end{vmatrix}}{-22} = \frac{8(-12) - (2(-7))}{-22} = -3.7$$

The solution to that system of equations is $(-2.27, -3.7)$. Whew. Your turn.

Now It's Your Turn

Find the determinant of each matrix. Find the answers in Appendix A.

19. $\begin{bmatrix} 4 & 3 \\ -9 & 1 \end{bmatrix}$

20. $\begin{bmatrix} 2 & -7 \\ -3 & -4 \end{bmatrix}$

21. $\begin{bmatrix} 11 & -11 \\ 7 & -9 \end{bmatrix}$

22. $\begin{bmatrix} 2 & 1 & 0 \\ -4 & -8 & 22 \\ 5 & -7 & 3 \end{bmatrix}$

23. $\begin{bmatrix} -10 & 1 & -1 \\ 3 & 8 & 12 \\ -4 & 2 & 0 \end{bmatrix}$

24. $\begin{bmatrix} 9 & -3 & 5 \\ 21 & -4 & -7 \\ 1 & 9 & 4 \end{bmatrix}$

Use Cramer's Rule to solve each linear system.

25. $6x + 15y = 6$

 $-2x + 4y = 20$

27. $5x + y = -20$

 $x - 5y = 11$

26. $2x + y = 2$

 $x + 2y = 8$

28. $6x - 8y = -30$

 $4x + 10y = 26$

An Introduction to Polynomials, Radicals, and Functions

4

Let's switch gears a bit. It is time to move beyond linear equations and get into some heavier algebra. Polynomials and their distant cousins the radicals will be the focus here. But by the time you work through this part, things such as $ax^2 + bx + c$ or $\sqrt{\dfrac{9}{81}}$ won't look like ancient hieroglyphics or some sort of communication from an alien in a distant galaxy.

If someone asked me what my least favorite part about algebra was, I'd have to say functions. But that is okay, you can't like everything (ahem—Brussels sprouts). And it is very possible that you disagree, think I am crazy, and love working through functions. If so, good for you. So let's work through this part together, step by step. The secret to functions is to make a lot of tables. Tables and the data that populate them will help you get through this and have a little *fun*ction at the same time.

Poly(nomial) Want a Cracker?

In This Chapter

- Adding, subtracting, and multiplying polynomials
- Factoring trinomials in the form of $x^2 + bx + c$ and $ax^2 + bx + c$
- Factoring polynomials of two squares and perfect squares
- Factoring polynomials completely using common binomials and grouping

This chapter will help prepare you for basic algebra problems, assignments in sciences such as chemistry and physics, and help you with more advanced math classes such as geometry or calculus. You might even become a more successful economist by mastering this chapter. Not bad for a single chapter in a book, is it? Polynomials, the topic of this chapter, have applications in all sorts of situations. Guess we'd better get started.

I'd Like You to Meet the Polynomials

This chapter spends time investigating the realm of polynomials. A polynomial is a monomial or a sum of monomials. That begs the question, "what is a monomial again?" A monomial is a number, a variable, or a product of a number and a variable that has whole numbers as the exponent. So a polynomial has one or more monomial, each called a term. Here is one:

$$3x^3 + x^2 - 4x + 16$$

This is a polynomial. It is composed of four different monomials: $3x^2$, x^2, $4x$, and 16.

One thing you have to look at is the degree of the monomial (and therefore the polynomial). The degree of the monomial is the exponent to which the variable is raised. The degree of $3x^3$ is 3. The degree of x^2 is 2. The degree of $4x$ is 1. And the degree of 16 is 0.

LOOK OUT!

The degree of a constant number (no variable) that is not zero is equal to 0. If the constant term is 0, there is no degree.

The degree of a polynomial is the greatest degree of all the terms.

Polynomials can have special names. Here is a short list of some of the polynomial names you may run across.

A monomial is a number or variable with whole number exponents.

A binomial is a polynomial with two terms.

A trinomial is a polynomial with three terms.

Adding Polynomials

Now that you are up to speed on polynomial-speak, let's get to work using them. Adding polynomials is no problem, as long as you remember to group like terms.

Example 1

Find the sum of these two polynomials:

$$(8x^3 + 4x^2 - x + 4) + (x^3 - 3x^2 + 4x - 12)$$

One way to do this is to arrange the like terms in each polynomial in a vertical formation:

$$
\begin{array}{rrrrr}
8x^3 & +\ 4x^2 & -\ x & +\ 4 \\
+\ x^3 & -\ 3x^2 & +\ 4x & -12 \\
\hline
9x^3 & +\ x^2 & +\ 3x & -8
\end{array}
$$

Another way to solve for the sum is to simply keep the equation horizontal and add the like terms that way:

$$(8x^3 + 4x^2 - x + 4) + (x^3 - 3x^2 + 4x - 12) =$$
$$(8x^3 + x^3) + (4x^2 - 3x^2) + (-x + 4x) + (4 - 12) =$$
$$9x^3 + x^2 + 3x - 8$$

Try your hands at subtracting polynomials. If you subtract, you want to add the opposite. Do this by multiplying all the terms of one of the polynomials by –1. Take a look at the following example.

Example 2

Find the difference of these two polynomials:

$$(5x^2 + 3x + 1) - (-6x^2 + 7x - 9)$$

But wait! You need to multiply one of the polynomials by –1 so you are adding it's opposite. Let's multiply the last polynomial by –1. Now the equation is …

$$
\begin{array}{r}
5x^2 + 3x + 1 \\
+\ \underline{6x^2 - 7x + 9} \\
11x^2 - 4x + 10
\end{array}
$$

Polynomials can be multiplied as well, and there are several different ways to approach multiplying two polynomials. The next examples look at two possible techniques.

Example 3

Multiply these polynomials using the distributive property:

$$(6y^2 + 3y + 2)(3y - 4)$$

Here you want to distribute the polynomial $3y - 4$ to each of the monomials in the other polynomial. It looks like this:

$$6y^2(3y - 4) + 3y(3y - 4) + 2(3y - 4)$$

Now multiply it out:

$$18y^3 - 24y^2 + 9y^2 - 12y + 6y - 8$$

Simplify by combining like terms:

$$18y^3 - 15y^2 - 6y - 8$$

There are a few tricks in algebra that would be helpful to learn. One of these tricks is a way to multiply binomials. This is done with the FOIL method.

PSST—TRY IT THIS WAY

The FOIL method is to multiply the **F**irst terms in each set of parentheses. Then multiply the **O**utside terms in each set of parentheses. Next multiply the **I**nside terms in each parentheses. And finally multiply the **L**ast terms in each set of parentheses. F-first O-outside I-inside L-last: FOIL.

Example 4

Use the FOIL method to find the product:

$$(4m + 6)(2m - 8)$$

First, multiply the first terms in each set of parentheses. That is $4m \times 2m = 8m^2$, and then multiply the outside terms in each set of parentheses. Or, $4m \times -8 = -32m$.

Next, multiply the inside terms. This is $6 \times 2m = 12m$. Finally, multiply the last terms in each parentheses: $6 \times -8 = -48$. Let's see what you are left with:

$$8m^2 - 32m + 12m - 48$$

Simplify to get $8m^2 - 20m - 48$. So, $(4m + 6)(2m - 8) = 8m^2 - 20m - 48$.

There are, of course, many different special polynomials. Let's take a look at how to solve a problem using one of those special cases.

Example 5

Find the product of the square of a binomial: $(x + 5)^2$

PSST—TRY IT THIS WAY

When asked to solve the square of a binomial, keep these two equations in mind:

$(a + b)^2 = a^2 + 2ab + b^2$

$(a - b)^2 = a^2 - 2ab + b^2$

Using the helpful formula given previously: $(x + 5) = x^2 + 2(5x) + 5^2$

Simplify to get the answer: $x^2 + 10x + 25$

Now It's Your Turn

Find the sum or difference of these polynomials. The answers are located in Appendix A.

1. $(4b^2 - 4) + (6b^2 + 7)$

2. $(3x - 5) - (3x^2 + 8x - 6)$

3. $(y^3 - 10) - (5y^3 + 7y^2 - 6y + 12)$

4. $(7r^3 + 2r - 15) + (5r^3 + 3r^2 + 17r - 18)$

Find the product of these polynomials.

5. $(5x - 2)^2$

8. $(8w - 3)(w + 1)$

6. $(2x^2 + 6x - 3)(4x - 2)$

9. $(5a + 9)(4a - 4)$

7. $(x^2 + 5x + 1)(3x + 2)$

Trinomials and Why We Love to Factor Them

Think of this section as the opposite of the previous section. In the previous section you were given two polynomials and asked to add, subtract, or multiply to get a final answer. In this section, you are given the final answer and are asked to find the polynomials that were manipulated to get that final form.

Let's start this part of our adventure by looking at how to factor out the GCF. What we are looking for here are values that are common to each term in the polynomial.

Example 6

Factor $6x^3y^2 - 18x^2y^3 + 9xy^2$

Here we have three terms. Each term is divisible by 3 and each term also contains at least one x and two y values. So the GCF is $3xy^2$. So all we do here is put the GCF term in front and group the others together after we divide each by the GCF like so:

$3xy^2(2x^2 - 6xy + 3)$

You can always check your answer by distributing the GCF back to each term.

Next we will move to trinomials in the form $x^2 + bx + c$.

The generalized formula for factoring trinomials is …

$$x^2 + bx + c = (x + n)(x + m) \text{ as long as } n + m = b \text{ and } nm = c$$

It is important (and necessary) to keep the signs in mind as you go through this process. Check out the table to find some helpful hints.

b	c	m	n
+	+	+	+
−	+	−	−

Note that when c is negative, then m and n have different signs.

Example 7

$x^2 + 8x + 12$

What you are looking for here are two positive factors of 12 that add up to 8. One way to do this is to make a table.

Factors of 12	Sum of Those Factors
12, 1	12 + 1 = 13 {Nope, not the right answer.}
2, 6	2 + 6 = 8 {Yeah! This is it.}
3, 4	3 + 4 = 7 {Nope, not the right answer.}

So this means that 2 and 6 are the values of m and n:

$$x^2 + 8x + 12 = (x + 2)\ (x + 6)$$

Check your answer by multiplying …

$$(x + 2)(x + 6) = x^2 + 8x + 12$$

Example 8

$x^2 + 3y - 18$

Watch out for the signs here. In this case, c is negative. Check back at the table and you will see that this means that m and n have to have opposite signs. Just keep that in mind as you work through this example. You are looking for the factors of −18, which have a sum of 3.

Factors of −18	Sum of Those Factors
−18, 1	−18 + 1 = −17 {no}
18, −1	18 + −1 = 17 {no}
−2, 9	−2 + 9 = 7 {no}
2, −9	2 + −9 = −7 {no}
−3, 6	−3 + 6 = 3 {yes!}
3, −6	3 + −6 = −3 {no}

This means that $x^2 + 3y - 18 = (x - 3)(x + 6)$.

Factoring Trinomials in the Form *ax²* + *bx* + *c*

This process is going to be the same as before. Watch your signs!

LOOK OUT!

Not only do you have to keep track of the signs for b, c, m, and n but you also have to keep track of the sign for a. It is very helpful to make a little chart to keep track of the factors and their signs!

Example 9

Factor this trinomial:

$$2x^2 + 17x + 8$$

All the factors are positive here so the signs are all positive (thank goodness for that)!

Not only do you have to be aware of the factors of 8, but you also need to take into consideration the factors of 2. A table can most definitely help. Keep in mind that you want to find the factors that have a sum of 17.

Factors of 2	Factors of 8	Possible Factorization	Middle Term
1, 2	1, 8	$(x + 1)(2x + 8)$	$10x$
1, 2	8, 1	$(x + 8)(2x + 1)$	$17x$
1, 2	2, 4	$(x + 2)(2x + 4)$	$8x$
1, 2	4, 2	$(x + 4)(2x + 2)$	$10x$

The correct factorization for the trinomial is $(x + 8)(2x + 1)$. So $2x^2 + 17x + 8 = (x + 8)(2x + 1)$.

Now It's Your Turn

Factor each of the following trinomials in the form $x^2 + bx + c$.

You can find the answers in Appendix A.

10. $t^2 - t - 20$

12. $w^2 - 2w - 24$

11. $x^2 - 10x + 21$

Factor each of the following trinomials in the form $ax^2 + bx + c$.

13. $3m^2 - 2m - 5$

15. $-3g^2 + 12g - 9$

14. $4x^2 + 27x - 7$

Factoring the Special Polynomials

There are several special polynomials that involve squares. Let's take a look at how they are factored.

Example 10

Factor these polynomials:

$$y^2 - 49$$

> **PSST—TRY IT THIS WAY**
>
> Keep these equations in mind when looking at factoring squares:
>
> $a^2 - b^2 = (a+b)(a-b)$ $\qquad a^2 + 2ab + b^2 = (a+b)^2$ $\qquad a^2 - 2ab + b^2 = (a-b)^2$

In this case, y^2 and 49 are both squares. In other words, this polynomial could be written as $a^2 - b^2$. So factoring this polynomial would look like this:

$$y^2 - 49 = y^2 - 7^2$$
$$= (y + 7)(y - 7)$$

Here is another example for you.

$$n^2 - 16n + 64$$

This is an example of a perfect square trinomial. Factor it by writing it in the form $a^2 - 2ab + b^2$:

$$n^2 - 2(8n) + 8^2$$
$$= (n - 8)^2$$

Now It's Your Turn

Factor these special polynomials. The answers are in Appendix A.

16. $4y^2 - 49$

17. $9w^2 - 81$

18. $16r^2 + 8rs + s^2$

19. $t^2 + 6t + 9$

20. $2p^2 - 20p + 50$

Factoring Polynomials Completely

There always seems to be one more thing that can be done with polynomials and this is no exception. This section walks through how to completely finish off the polynomial.

Factoring Using Binomials

One way to completely factor a polynomial is to factor out the binomials. Look at the following example.

Example 11

Factor this expression using a common binomial:

$$3x(x + 8) - 4(x + 8)$$

Both polynomials contain the binomial $(x + 8)$. Factor that out of both of them and you are left with the following:

$$(x + 8)(3x - 4)$$

Sometimes the common binomial doesn't just jump out at you so you want to find another way. There is another way to factor polynomials completely—factor by grouping!

Example 12

Factor this polynomial completely by grouping:

$$x^3 + 6x^2 + 3x + 18$$

The first thing to do here is to factor the common monomial from pairs of terms. Taking a closer look here, it appears as if there is a common monomial of $(x + 6)$ in there. Splitting the whole polynomial into two sets of pairs may help see this. Let's group the polynomial like this:

$$(x^3 + 6x^2) + (3x + 18)$$

Now, see how there is the monomial $(x + 6)$ in each pair?

$$x^2(x + 6) + 3(x + 6)$$

Use the distributive property to rewrite this to read …

$$(x + 6)(x^2 + 3)$$

This is what the polynomial looks like when completely factored.

Now It's Your Turn

Use grouping or common binomials to factor each polynomial completely. You can find the answers in Appendix A.

21. $x^3 + 4x^2 + x + 4$

22. $a^2 + 5b + ab + 5a$

23. $7y^2(y - 4) + 6(4 - y)$

24. $r^2 - 6r + 9$

25. $3y^3 + 9y^2 - 12y$

26. $ab - 10b - 2a + 20$

27. $4y^3 + 16y^2 + 32y$

28. $x^3 - 49x$

29. $8x^3 - 4x^2y - 32y + 16$

30. $x^2 + 5x - 24$

Totally Radical

In This Chapter

- How a radical looks in its simplest form
- How to simplify radicals when you are asked to add or subtract
- What the product and quotient properties are of radicals
- How to solve radical equations in two simple steps

It is time to rebel, to get a little radical. Ok, maybe not too rebellious but it is time to look at radicals. These are things like square roots that sneak in many algebra or scientific problems. So, go ahead, give yourself permission to feel a little wild. Sharpen your pencils and let's learn about radicals.

Dig into Your Roots

Know those crazy square root signs in algebra ($\sqrt{}$) that have numbers inside and/or outside the square root? These are collectively known as radicals. The number inside the radical is the radicand.

Radical expressions contain at least one variable in the radicand. For example $\sqrt{2x}$ is a radical expression. So is $\sqrt[3]{4x-3}$. Not pretty is it? But we will work through these step by step.

There are three simple rules to follow when working to simplify a radical:

1. Make sure there are no perfect square factors in the radicand (except for 1, of course).

2. Make sure there are no fractions in the radicand.

3. Work to ensure that there are no radicals in any denominator.

Sometimes it's easier to do something than just talk about it.

Example 1

Simplify the following expression by looking for a perfect square:
$$\sqrt{100}$$

Yikes. But if you step back and factor 100, you'll find that it is equal to two numbers that are perfect squares:

$$\sqrt{100} = \sqrt{25 \times 4}$$

This is equal to

$$\sqrt{25} \times \sqrt{4}$$

$$5 \times 2 = 10$$

PSST—TRY IT THIS WAY

Knowing a handful of perfect squares makes your life easier (and impresses friends and relatives). Keep these in mind:

$$\sqrt{4} = 2 \quad \sqrt{9} = 3 \quad \sqrt{16} = 4 \quad \sqrt{25} = 5 \quad \sqrt{36} = 6$$

$$\sqrt{49} = 7 \quad \sqrt{64} = 8 \quad \sqrt{81} = 9 \quad \sqrt{100} = 10$$

Example 2

Simplify $\sqrt{16x^3}$ using the *product property of radicals*.

See the two perfect squares in there? 16 is a perfect square because $4 \times 4 = 16$. And x^2 is a perfect square as well.

Rewrite this to break out the perfect squares:

$$\sqrt{16x^3} = \sqrt{16} \times \sqrt{x^2} \times \sqrt{x}$$

Simplify to get this:

$$\sqrt{16x^3} = 4x\sqrt{x}$$

DEFINITION

The **product property of radicals** says that the square root of a product is equal to the square root of the factors. And that is what you did in that example: $\sqrt{ab} = \sqrt{a} \times \sqrt{b}$.

PSST—TRY IT THIS WAY

You may have guessed but there is also a quotient property of radicals as well. This says that the square root of a quotient is equal to the square root of the numerator and the square root of the denominator.

In other words, $\sqrt{\dfrac{a}{b}} = \dfrac{\sqrt{a}}{\sqrt{b}}$ as long as $a \geq 0$ and $b > 0$.

Example 3

Sometimes you are asked to simplify a radical with a fraction.

For example, simplify $\dfrac{2}{\sqrt{11}}$.

You cannot have a radical in the denominator. To get rid of it, simply multiply by 1. It's just that in this case, 1 is in the form $\dfrac{\sqrt{11}}{\sqrt{11}}$.

When $\dfrac{2}{\sqrt{11}}$ is multiplied by $\dfrac{\sqrt{11}}{\sqrt{11}}$, use the product property of radicals.

This is equal to $\dfrac{2\sqrt{11}}{\sqrt{121}}$.

Simplify to get $\dfrac{2\sqrt{11}}{11}$.

What you just did has a special name. It is called *rationalizing the denominator.*

DEFINITION

The process of ridding the denominator of a radical is called **rationalizing the denominator.**

Now It's Your Turn

Simplify the following radical expressions. The answers can be found in Appendix A.

1. $\dfrac{1}{\sqrt{5}}$

2. $\sqrt{\dfrac{9}{81}}$

3. $\sqrt{24}$

4. $3\sqrt{12}$

5. $\dfrac{4}{\sqrt{7}}$

Simplifying Radicals and Adding

Adding radicals will probably involve the distributive property. Take a look at the next example.

Example 4

Add these two radicals:

$$8\sqrt{5} + \sqrt{45}$$

Notice that there is one perfect square in $\sqrt{45}$. It is equal to $\sqrt{9 \times 5}$.

Now the equation looks like this:

$$8\sqrt{5} + \sqrt{9} \times \sqrt{5}$$

Simplify $8\sqrt{5} + 3\sqrt{5}$.

Now add them and you'll find that $8\sqrt{5} + \sqrt{45} = 11\sqrt{5}$. Subtracting radicals is similar to this, as you learn in the next section.

Simply Subtracting

When subtracting radicals, factoring out perfect squares can work wonders.

Example 5

Subtract these two radicals:

$$6\sqrt{3} - 2\sqrt{12}$$

Notice the perfect square in there.

Simplify these:

$$6\sqrt{3} - 2\sqrt{4 \times 3}$$
$$6\sqrt{3} - 2(2 \times \sqrt{3})$$

Keep going:

$$6\sqrt{3} - 4\sqrt{3}$$

Subtract to get the following:

$$6\sqrt{3} - 2\sqrt{12} = 2\sqrt{3}$$

Multiplying Radicals

When multiplying radicals, it is necessary to use the product property of radicals. And guess what? You have already been using that and you didn't even realize it.

Try your hand at an example.

Example 6

Find the product using the product property of radicals:

$$\sqrt{5xy^2} \times 7\sqrt{x}$$

First, multiply the radicals to get $7\sqrt{5x^2y^2}$.

Check for the square roots of the factors inside the radicand. See any? Rewrite the radical:

$$7 \times \sqrt{5} \times \sqrt{x^2} \times \sqrt{y^2}$$

Simplify to get rid of any perfect squares and this gives the answer that follows:

$$7xy\sqrt{5}$$

Dividing Those Radicals

When dividing radicals it is important to keep the quotient property of radicals in mind.

LOOK OUT!

Remember those rules for dealing with radicals that were presented at the beginning of this chapter? One of them is that when dividing radicals, work to ensure that there are no radicals in any denominator!

Example 7

Find the quotient using the quotient property of radicals:

$$\sqrt{\frac{3}{r^2}}$$

According to the quotient property of radicals this is equal to this:

$$\frac{\sqrt{3}}{\sqrt{r^2}}$$

Simplify to get rid of the square roots that might be hanging out in the quotient.

The final answer is $\dfrac{\sqrt{3}}{r}$.

That is simple enough, but what if the radical in the denominator is not a perfect square? How can you work with that so you are able to remove the radical? Try a little rationalization.

Example 8

Find the quotient:

$$\frac{\sqrt{7}}{\sqrt{5x}}$$

So here's the problem. You need to get rid of the radical in the denominator but $\sqrt{5x}$ is not a perfect square (it's not a perfect anything for that matter). So, like a Fairy Godmother, you'll have to make it a perfect square, if only for a moment.

Multiply the numerator and the denominator of the quotient by $\sqrt{5x}$:

$$\frac{\sqrt{7}}{\sqrt{5x}} \times \frac{\sqrt{5x}}{\sqrt{5x}}$$

LOOK OUT!

This looks extreme doesn't it? But you really haven't changed anything. Multiplying the numerator and denominator by the same number is actually just multiplying the number by 1.

Now use the product property of radicals:

$$\frac{\sqrt{35x}}{\sqrt{25x^2}}$$

Look at that wonderful perfect square in the denominator. Now you can get rid of it.

$$\frac{\sqrt{35x}}{\sqrt{25} \times \sqrt{x^2}} = \frac{\sqrt{35x}}{5x}$$

Dividing radicals is a little magical. The denominator is changed into a perfect square and then changed back. As long as nothing is turned into a toad, you are on the right track.

Now It's Your Turn

Add or subtract the radicals to simplify the following. You can find the answers in Appendix A.

6. $2\sqrt{3} + 7\sqrt{3}$

8. $5\sqrt{6} - 3\sqrt{54}$

7. $2\sqrt{3} - 7\sqrt{3}$

9. $4\sqrt{4} + 6\sqrt{48}$

10. $\sqrt{5} + 9\sqrt{50}$

12. $3\sqrt{12} + 2\sqrt{2}$

11. $7\sqrt{6} - \sqrt{72}$

13. $\sqrt{18} + 3\sqrt{3}$

Multiply or divide the radicals to simplify the following:

14. $\sqrt{5} \times \sqrt{45}$

17. $\sqrt{\dfrac{x}{45}}$

15. $\sqrt{8}(6 - \sqrt{8})$

18. $\sqrt{\dfrac{4}{y^2}}$

16. $(\sqrt{5} + \sqrt{7})(4 + \sqrt{8})$

19. $\dfrac{2}{\sqrt{15}}$

Solving Those Radical Equations

Sometimes you will be faced with an equation that has a variable in the radicand. Those are called radical equations. There are two steps to solving these equations. Master these and you'll be off and running with the rest of this chapter. The first step is to isolate the radical on one side of the equation. Then square both sides of the equation. Problem solved.

The following section shows an example.

Example 9

Solve this radical equation:

$$\sqrt{y} - 11 = 0$$

First thing to do is to isolate the radical on one side of the equation. To do that in this example, add 11 to each side. The equation will now read as follows:

$$\sqrt{y} = 11$$

Square each side of the equation:

$$(\sqrt{y})^2 = 11^2$$

And solve:

$$y = 121$$

Problem solved. Here is one more example before it's your turn.

Example 10

Solve this equation (look out—there may be a twist to this one):

$$\sqrt{8 + 2x} = x$$

Okay, this one looks a little more complicated than the previous example. But walk through the steps. The first thing to do is to isolate the radical on one side of the equation. That's already been done.

Now square each side:

$$(\sqrt{8 + 2x})^2 = x^2$$
$$8 + 2x = x^2$$

Simplify this equation and write it in standard form:

$$0 = x^2 - 2x - 8$$

Factor:

$$0 = (x - 4)(x + 2)$$

Now solve for x; there are two possible answers:

$$x = -2 \text{ or } x = 4$$

Check these answers back in the original equation to see if they work.

If you plug –2 into	If you plug x into
$\sqrt{8+2x} = x$	$\sqrt{8+2x} = x$
it is equal to	it is equal to
$\sqrt{8+2(-2)} = -2$	$\sqrt{8+2(4)} = 4$
$2 = -2$	$4 = 4$
That isn't correct!	That is correct!

The only actual solution for this equation is 4. –2 is not a solution. It is called an *extraneous solution*.

DEFINITION

Extraneous solutions are answers that are derived from the squaring of both sides of an equation. These solutions do not hold true for the original equation before it was squared. Be sure to check all your answers back into the original equation to see if they are extraneous or not.

Now It's Your Turn

Solve each radical equation. Be on the lookout for extraneous solutions! The answers are located in Appendix A.

20. $5\sqrt{x} - 10 = 0$

21. $\sqrt{5x} + 4 = 16$

22. $4 = \sqrt{x-2}$

23. $\sqrt{5-4x} = x$

24. $\sqrt{x-11} = \sqrt{2x+6}$

25. $9\sqrt{y+3} - 8 = 0$

26. $\sqrt{18+7x} = 0$

27. $\sqrt{10x-25} = x$

28. $\sqrt{8x+16}=16$

29. $\sqrt{y-12}=y$

Functions— They Don't Begin with "Fun" for Nothing

In This Chapter

- Talking about functions: inputs and outputs
- Graphing functions using a table
- Determining whether a graph is of a function or not
- Flipping a function (or finding its inverse)
- Graphing a square root function

You know how sometimes you hear that something is a function of something else? The fact that a dentist discovers two new cavities in someone's teeth is most likely a function of that person either eating too much Halloween candy or not brushing and flossing enough. A function in algebra is just like that (but not as painful as a trip to have a cavity filled).

What Is the Function of a Function?

Algebraically speaking, a function has the following:

- A set of inputs and a set of outputs
- Each input is paired with only one output

The domain of a function is the inputs, and the range of the function is the outputs. Let's look at a few functions to get a better idea of what they are all about.

Example 1

The graph shows the ages of the top three finishers of the Kids Fun Run each year from 2004 to 2006. Does this set of data represent a function?

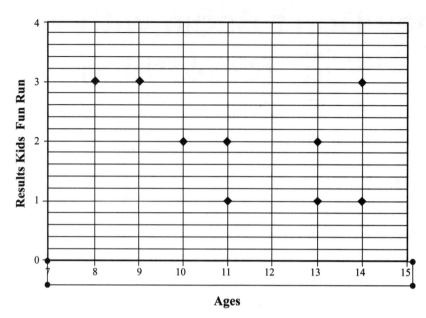

Graph results of Kids Fun Run.

Make a table to show the data in this graph. (Usually the table is given to you and you work that way. We are going to do it a little backward here to get going.)

Age	Place in Race
8 years	third
9 years	third
10 years	second
11 years	first
11 years	second
13 years	first
13 years	second
14 years	first
14 years	third

Remember that the definition of a function indicates that each input pairs with only one output. Is this a function? No. The data for 11 years has one kid coming in first place and one kid coming in second place. That input (11 years) pairs with more than one output (first and second place).

PSST—TRY IT THIS WAY

You will be graphing and looking at graphs of many functions. One way to tell if a certain graph is for an actual function is to do the vertical line test. Any given relation is a function if and only if any vertical line drawn on the graph hits only one spot. Try it, it works!

Example 2

Consider the following set of data. What is the range and domain of the set of data? Is this a function? Explain your answer.

(3,–4), (1,–2), (0,6), (–2,2), and (4,5)

If you are a visual learner or just someone who likes to see things in tables, you might want to place this data in a table.

x	y
3	–4
1	–2
0	6
–2	2
4	5

The range of a relation is the value of the outputs. In this case, the range is –4, –2, 2, 5, 6. The domain is the value of the inputs. In this problem, the domain is –2, 0, 1, 3, 4. Is this a function? Yes it is. Each input has only one output.

Now It's Your Turn

Find the range and domain of each set of numbers. Then indicate if each is a function or not. You can find the answers in Appendix A.

1. (3,10), (2,–1), (5,8), (–2,–1), (3,2)

2. (0,5), (1,6), (3,7), (8,1), (9,0)

3. (–4,2), (–1,5), (3,6), (6,2)

4. (–1,–1), (3,5), (4,8), (–2,0)

Graphing Functions

Most likely, you will be asked to graph functions. And to do so, it helps to make a table of possible input and output values. This is shown in Example 3.

Example 3

Graph the function $y = -4x + 2$.

First thing to do is make a table of values x and y. Choose a number to plot into x and then calculate out what the value of y would be.

x	y
0	2
−1	6
−2	10
1	−2
2	−6

Plot those points on a line.

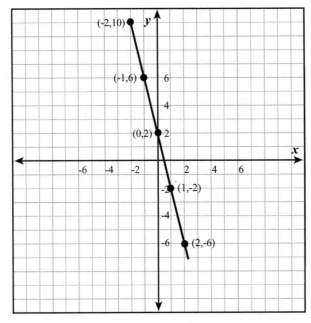

Graph of $y = -4x + 2$.

Double-check the data with the vertical line test. Notice how you can draw a vertical line through any of those points and none are represented by more than one point. It really *is* a function.

Try another example.

Example 4

Graph the function $y = \frac{1}{2}x + 3$.

Start with a table.

x	y
0	3
1	3.5
2	4
–1	2.5
–2	2

Now, graph those points.

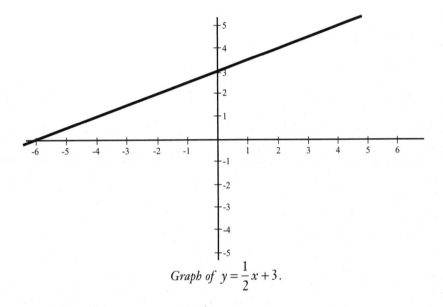

Graph of $y = \frac{1}{2}x + 3$.

Linear or Not?

A function is linear if it can be written in the form $y = mx + b$. The graph of a linear function is a line (as seen in the previous example).

If the y in $y = mx + b$ is renamed $f(x)$, the equation for that function is re-written in function notation. In function notation, the equation is $f(x) = mx + b$.

LOOK OUT!

The parentheses here in $f(x)$ are not, in this case, a way to signal multiplication. This is read "the value of f of x" or "f of x." This identifies x as the independent variable. The domain for this function is all the values of x for which $f(x)$ is defined. The range is the value of $f(x)$.

Example 5

Tell if the following function is linear:

$$f(x) = x^2 + 3x - 6$$

This is not linear. The equation is not in the form $f(x) = mx + b$.

Tell if the following function is linear:

$$f(x) = 7x + 19$$

This is linear. The equation is in the form of $f(x) = mx + b$.

Now It's Your Turn

Graph the following functions by making a table of values first. The answers are located in Appendix A.

5. $y = -x + 10$

7. $y = \dfrac{2}{5}x + 1$

6. $y = 4x - 8$

8. $y = -8$

Tell if each function is linear.

9. $f(x) = x + 19$

10. $f(x) = x^3 + 5x^2 + 8x - 4$

Special Functions

There are a couple of special cases that we need to cover: inverse and square root functions.

Inverse Functions

A regular function has a domain and a range. But to find its inverse is a whole different story. Think of the inverse of a function as its alter ego. You will actually switch the x and the y in the equation, getting a whole new domain and range.

Start small.

Example 6

Find the inverse of the equation $y = 3x + 7$.

The original expression is $y = 3x + 7$.

To find the inverse, switch the x and the y in the equation. Now the equation looks like this:

$$x = 3y + 7$$

Yet again, however, you want to get the y all by itself on a side. So subtract 7 from each side:

$$x - 7 = 3y$$

Solve for y:

$$\frac{1}{3}x + \frac{7}{3} = y$$

This is the inverse of the original equation $y = 3x + 7$.

That was an example with a regular equation. What does the inverse of a function look like? Two functions are inverses of each other if …

$$f(g(x)) = x \text{ and } g(f(x)) = x$$

The function g is usually written as f^{-1}. When speaking, this is stated as "f inverse." Put this new knowledge to work for you.

Example 7

Find the inverse for this function:

$$f(x) = \frac{x+2}{5}$$

As an equation, this is originally $y = \frac{x+2}{5}$.

Switch the places of the y and x (you are looking for the inverse after all). Now you have this:

$$x = \frac{y+2}{5}$$

The idea is to get y by itself. Multiply each side by 5:

$$5x = y + 2$$

Subtract 2 from each side to get y alone:

$$5x - 2 = y$$

Or:

$$g(x) = 5x - 2$$

REALITY CHECK

When would this knowledge of finding the inverse of a function come in handy in everyday life? Suppose your thermometer at home reads only the temperature in degrees Celsius. To convert between Celsius and Fahrenheit, you would use an inverse function.

Now It's Your Turn

Find the inverse of each function. You can find the answers in Appendix A.

11. $y = 4x + 5$

14. $y = \frac{x+5}{3}$

12. $y = -5x - 1$

15. $f(x) = x + 6$

13. $y = \frac{3}{x}$

16. $f(x) = 5 - \frac{1}{3}x$

17. $f(x) = \dfrac{13}{x-2}$

19. $f(r) = 4 + \dfrac{2}{3}x$

18. $f(x) = 6x + 1$

20. $f(x) = -x$

Square Root Functions

The best, easiest, and recommended way to determine the range and domain of a function is to set up a table. This is true for functions that contain a square root, too. Square root functions take the form $y = a\sqrt{x}$.

Example 8

Find the domain and the range of the function $y = \sqrt{x}$. First thing's first. Set up a table of possible values for x and y. Plug values in for x and find their corresponding value of y.

x	y
1	$\sqrt{1}$
2	$\sqrt{2}$
4	2
9	3

LOOK OUT!

Remember that the square root of a negative number is undefined. Any table you set up to work with a function of a square root is going to be ≥ 0.

Remember that the domain of the function is the values for x and the range is the values for y.

The next logical step when faced with a function and values for the domain and range is to graph it. Use the domain and range values for $y = \sqrt{x}$ to graph this function.

Example 9

Graph the function $y = \sqrt{x}$.

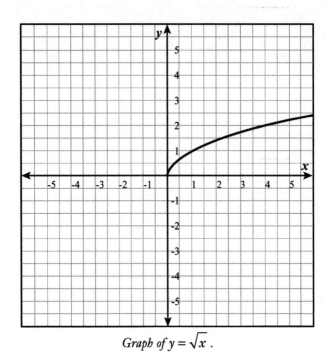

Graph of $y = \sqrt{x}$.

Let's do another example.

Example 10

Find the domain and the range of the function $y = 4\sqrt{x} + 3$. Then graph that function. First make the table for the function. Substitute values for x into the table and then solve to get the corresponding value for y.

x	y
0	3
1	7
2	8.6
4	11
9	15

Now graph it.

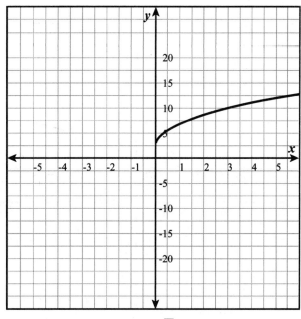

Graph of $y = 4\sqrt{x} + 3$.

PSST—TRY IT THIS WAY

Notice that when making the table for the domain and the range, the perfect square is chosen to make the math simpler. That is okay. You can choose any number you want but keep in mind that although your axes may be bigger, it is often easier to calculate whole numbers.

These functions can look like just about anything. Take a look at the function in the next example.

Example 11

Find the domain and range of the function $y = 3\sqrt{x-4} + 2$. Then graph it.

Don't panic. There is actually a shortcut to doing this problem (how often can you say that?).

This function can also be written in the format $y = a\sqrt{x-b} + k$.

In this particular case, $b = 4$ and $k = 2$. File this information for later.

When you get rid of the b and the k from the equation you are left simply with $y = a\sqrt{x}$. Or in this equation, $y = 3\sqrt{x}$. Determine the range and the domain of that function.

x	y
0	0
1	1
4	6
9	9

Now graph it.

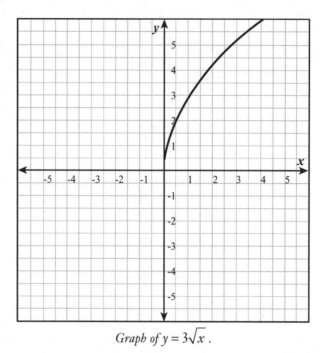

Graph of $y = 3\sqrt{x}$.

But wait! What about those values of h and k? Remember that h is equal to 4 and k is equal to 2. What this means is that the entire graph you have just made needs to be shifted to the right 4 units and up 2 units. The final graph of $y = 3\sqrt{x - 4} + 2$ looks like this:

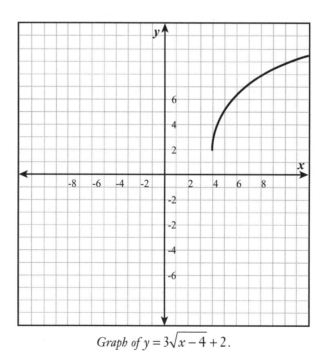

Graph of $y = 3\sqrt{x - 4} + 2$.

Now It's Your Turn

Find the domain and range of each function, and then graph the function. The answers are located in Appendix A.

21. $y = 3\sqrt{x}$

22. $y = \frac{1}{2}\sqrt{x}$

23. $y = -6\sqrt{x}$

24. $y = \frac{1}{4}\sqrt{x}$

25. $y = -0.6\sqrt{x}$

26. $y = \sqrt{x} + 8$

27. $y = \sqrt{x} - 1$

28. $y = \sqrt{x+2} - 3$

29. $y = \sqrt{x+2} + 4$

30. $y = -\sqrt{x+1} - 5$

Quadratic Everything

Now, this stuff I like—the quadratic equation $y = ax^2 + bx + c$ and all its varied uses and permutations. Not only can this be useful later in geometry classes, but if you like to solve math problems by regular ordered steps, then this is the part for you. Time to solve and graph a few quadratic equations.

Quadratic Equations

In This Chapter

- Graphing quadratic functions in the form $y = ax^2 + c$ and in the form $y = ax^2 + bx + c$
- Using graphs to solve quadratic equations
- Using square roots or completing the square to solve quadratic equations
- Solving quadratic equations using the quadratic formula
- Interpreting discriminants to find information about the number of solutions for a particular quadratic equation

The quadratic equation is one of those things you should have committed to memory. Participation in television quiz shows, dinner party conversations, and of course algebra problems, are all times when knowing this might come in handy.

Quadratic Functions in Standard Form

The quadratic equation is any equation that can be written in the form:

$ax^2 + bx + c = 0$ as long as $a \neq 0$

Quadratic functions are functions that can be written in that form. The shape of the graph of a quadratic function is a *parabola*.

DEFINITION

A **parabola** is a U-shaped graph. The graph of a quadratic function has this general shape.

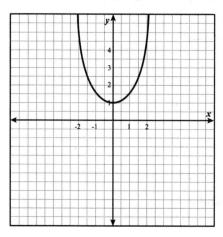

Generalized parabolic graph.

The graph against which all other graphs of quadratic functions is compared is called the parent quadratic function. This is $y = x^2$. Refer back to this graph each time you graph one of the quadratic functions in this chapter. It is the graph of the most basic quadratic function.

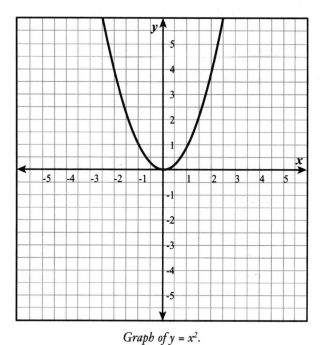

Graph of $y = x^2$.

A good place to start is to work with equations where $b = 0$. This leaves the quadratic function as $y = ax^2 + c$.

Example 1

Graph the equation $y = 4x^2$ (notice in this equation that b and c both equal 0).

Make a table of possible values.

x	y
0	0
−1	4
−2	16
1	4
2	16

Now make a graph using the numbers in the table.

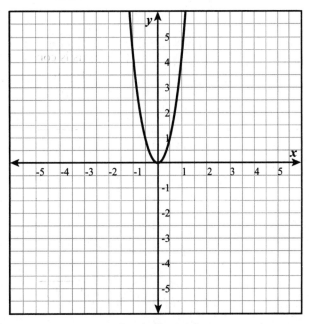

Graph of $y = 4x^2$.

Compare this graph with that of the parent quadratic function. Notice how the new graph is thinner than the graph for $y = x^2$ and seems to fit inside it? The graph of $y = 4x^2$ has been stretched vertically by a factor of 4.

Try another one.

Example 2

$$y = -\frac{1}{2}x^2$$

Again, make a table of the possible values.

x	y
0	0
–1	$-\dfrac{1}{2}$
–2	–2
–3	$-\dfrac{9}{2}$
–4	–8
1	$-\dfrac{1}{2}$
2	–2
3	$-\dfrac{9}{2}$
4	–8

Before you look at the graph, what do you notice about the numbers in the table? They are all negative numbers aren't they? One would expect that the graph would lie below the x-axis.

And it does. Here is the graph.

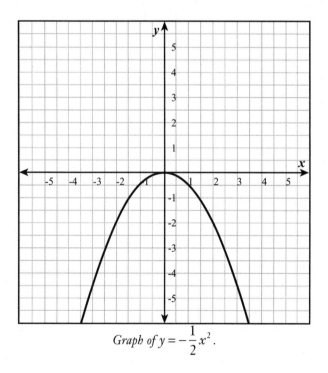

Graph of $y = -\dfrac{1}{2}x^2$.

PSST—TRY IT THIS WAY

Notice anything with the two examples you just worked through? When the value of a in the quadratic function is positive, the graph opens upward. When the value of a is negative, the graph opens down. Good things to keep in mind as you work through the examples.

Let's take a look at a few more examples.

Example 3

Graph:

$$y = \frac{1}{3}x^2 + 2$$

This is graphing the equation in the form $y = ax^2 + c$.

Make the table of values.

x	y
0	2
−1	$2\frac{1}{3}$
−2	$3\frac{1}{3}$
−3	5
1	$2\frac{1}{3}$
2	$3\frac{1}{3}$
3	5

Now graph the equation.

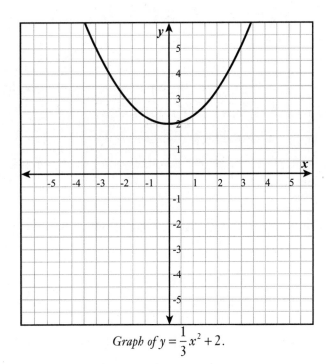

Graph of $y = \dfrac{1}{3}x^2 + 2$.

This graph has moved up the y-axis two spaces. You can reason that when the value of c is positive, the graph will move up the y-axis. When the value of c is negative, the graph is located farther down on the y-axis.

Now It's Your Turn

Graph the following. The answers can be found in Appendix A.

1. $y = 7x^2$

2. $y = x^2 - 3$

3. $y = x^2 - 8$

4. $y = x^2 - 9$

5. $y = \dfrac{13}{2}x^2$

Graphing in the Form $y = ax^2 + bx + c$

You are now very good at graphing simple quadratic functions. It's time to take it one step further and graph a generalized form of quadratic function. And that means it's time to graph $y = ax^2 + bx + c$.

There are rules and properties that you need to keep in mind as you work through these problems:

- The graph has an axis of symmetry equal to $x = -\dfrac{b}{2a}$.
- The vertex of the graph has the x-coordinate $-\dfrac{b}{2a}$.
- The y-intercept point is c. You can check to make sure the graph is correct because $(0,c)$ should lie on the parabola.

Let's put those rules to work for us as you work through an example.

Example 4

Graph $y = 2x^2 - 4x + 3$.

Look at the equation. You can tell just by looking at it whether the parabola that represents the equation opens up or down. The value of a is positive so the parabola opens up.

LOOK OUT!

The graph of the quadratic equation is a parabola that opens upward if the value of $a > 0$. The parabola opens downward if the value of $a < 0$.

The next thing you should do when graphing a quadratic equation in the form $y = ax^2 + bx + c$ is to determine the axis of symmetry.

Remember that the axis of symmetry is $x = -\dfrac{b}{2a}$.

In this particular case:

$$x = -\dfrac{-4}{2(2)} \text{ or } x = \dfrac{1}{1} = 1$$

Plot this on your graph. Then, find the vertex of the graph. Remember that the x coordinate of the vertex is at $-\dfrac{b}{2a}$ which you already know is 1. Find the y coordinate and add the value of x into the equation:

$$2x^2 - 4x + 3 = 2(1)^2 - 4(1) + 3$$
$$= 2 - 4 + 3 = 1$$

So the coordinates of the vertex are (1,1).

Choose x values so that you find two points to plot on each side of the vertex. Then find the corresponding y values. The points can be (0,3) (-1,9) to the left, and (2,3) (4,19) to the right.

Plot the points on the graph. Then reflect the points across the axis of symmetry and draw the parabola. The final result should look like this:

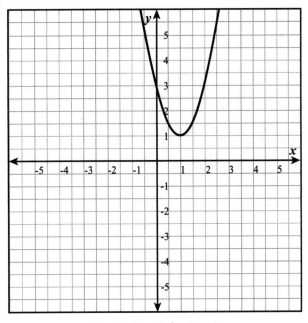

Graph of $y = 2x^2 - 4x + 3$.

Try another example. There are a lot of steps here.

Example 5

Find the axis of symmetry and the vertex of this equation. Graph the final result.

$x^2 + 2x - 1$

The axis of symmetry, $x = -\dfrac{b}{2a}$

$= -\dfrac{2}{2(1)} = -1$

The x value of the vertex is –1. The y value of the vertex is equal to this:

$x^2 + 2x - 1$

$(-1)^2 + 2(-1) - 1 = -2$

The vertex is (–1,–2).

Plot two points: (0,–1) and (2,1).

Reflect those points across the axis of symmetry and graph the parabola. The final result should look like this:

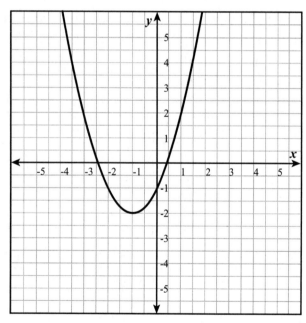

Graph of $y = x^2 + 2x - 1$.

Now It's Your Turn

Find the axis of symmetry and the vertex of each equation. Then graph the results. The answers are located in Appendix A.

6. $y = 2x^2 - 3x + 9$

9. $y = 9x^2 + 9x$

7. $y = -x^2 + 4x + 12$

10. $y = \frac{1}{3}x^2 + 3x - 5$

8. $y = -3x^2 + 12x + 4$

Solving Quadratic Equations

Now that you have learned to graph quadratic functions, it's time to move to solving actual problems using the quadratic equation. Remember that the quadratic equation is $ax^2 + bx + c = 0$. There are several ways to solve quadratic equations. Because graphing is fresh in your mind, let's start with that method.

Solving Quadratic Equations by Graphing

There are a few steps to follow here. First make sure that the equation is in the form $ax^2 + bx + c = 0$. Rewrite it if necessary.

Then, graph the function that is related to that equation. The function will have the general form $y = ax^2 + bx + c$.

The places where the graph crosses the x-axis are the solutions for $ax^2 + bx + c = 0$.

Example 6

Solve $x^2 + 2x = 8$ by graphing.

First, write the equation in the form $ax^2 + bx + c = 0$:

$$x^2 + 2x - 8 = 0$$

Graph the corresponding function $y = x^2 + 2x - 8$.

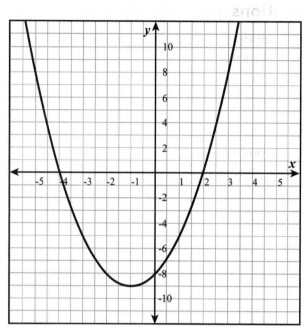

Graph of y = x² + 2x − 8.

Take note of where the graph crosses the *x*-axis. It crosses at two points: (−4,0) and (2,0).

The *x*-intercepts of −4 and 2 are solutions.

Check these answers by subbing the value for *x* back into the original equation.

For $x = -4$	For $x = 2$
$x^2 + 2x = 8$	$x^2 + 2x = 8$
$(-4)^2 + 2(-4) = 8$	$(2)^2 + 2(2) = 8$
$8 = 8$	$8 = 8$

They both check out!

PSST—TRY IT THIS WAY

Not all quadratic equations have two solutions. There is a shortcut to know how many solutions you are looking for. It all depends on how many times the graph of the equation crosses the *x*-axis.

- There are two solutions if the graph crosses the *x*-axis twice, having two *x*-intercepts.
- It is possible to "cross" the *x*-axis just once with a parabola. It can "touch" it once.
- There are no real solutions if the graph does not cross the *x*-axis at all.

Another way to solve quadratic questions is to use square roots.

Solving Quadratic Equations with Square Roots

Suppose you are faced with a quadratic equation that is in the form $ax^2 + c = 0$. If that is the case, you should consider solving it using square roots.

Here are the steps to doing this:

a. Rearrange so that x^2 is alone on one side of the equation. This will put the equation in the form $x^2 = d$.

b. If $d > 0$ then there are two solutions. That is $x = \pm\sqrt{d}$.

c. If $d = 0$ then there is one solution. That solution is $x = 0$.

d. If $d < 0$ then there is no solution.

The following section shows an example.

Example 7

Solve this using square roots:

$$x^2 + 10 = 4$$

Rearrange so that this is in the form $x^2 = d$.

$$x^2 = -6$$

This won't work. Negative numbers do not have a real square root. There is no solution. You need to work through another example.

Example 8

Solve using square roots:

$$3x^2 = 27$$

Divide both sides by 3 to get the equation in the form $x^2 = d$:

$$x^2 = 9$$

Solve for x:

$$x = \pm\sqrt{9} = \pm 3$$

The solutions to this quadratic equation are 3 and –3.

Squares, squares everywhere. You can also solve a quadratic equation by completing the square.

Completing Squares to Solve Quadratic Equations

When faced with a quadratic equation in the form $x^2 + bx$, you can actually add a constant to it. This makes the expression $x^2 + bx + c$. This is what is known as completing the square.

When you begin a problem where you are going to solve the quadratic equation by completing the square, be sure to rearrange so that the equation is in the form $x^2 + bx = d$.

PSST—TRY IT THIS WAY

In plain English, when completing the square to solve a quadratic formula, you should add the square of half the coefficient (b) to the term bx. This looks like $x^2 + bx + (\frac{b}{2})^2$. This can be rewritten as $(x + \frac{b}{2})^2$.

Example 9

Solve this by completing the square:

$$3x^2 + 24x - 9 = 0$$

Rearrange the equation so that it is in the form $x^2 + bx = d$. Do this by adding 9 to each side.

$$3x^2 + 24x = 9$$

Remember that you really want this in the form of $x^2 + bx = d$. So divide each side by 3. Now the equation is this:

$$x^2 + 8x = 3$$

Now it is clear that $b = 8$. To work to complete the square you need to add the square of one half of b to both sides. In this case, you add 4^2 to each side:

$$x^2 + 8x + 16 = 3 + 16$$

Factor the binomial on the left side:

$$(x + 4)^2 = 19$$

Take the square root of both sides:

$$x + 4 = \pm\sqrt{19}$$

Solve for x:

$$x = -4 \pm \sqrt{19}$$

So the solutions to this quadratic equation are 0.35 and -8.35. These are not the neatest solutions but algebra is not always neat and tidy.

Using the Quadratic Formula to Solve the Quadratic Equation

Whoa. Did you have to shake your head and read that again? It is possible to solve a quadric equation using the quadratic formula.

The quadratic formula is $x = \dfrac{-b \pm \sqrt{b^2 - 4ac}}{2a}$.

Sometimes when solving a quadratic equation it is just as easy to use the quadratic formula to find the answer as to use any other method.

Example 10

Solve the equation:

$$3x^2 - 6 = x$$

First step is to write this equation in standard form ($ax^2 + bx + c = 0$):

$$3x^2 - x - 6 = 0$$

If you were looking for the values of a, b, and c (which you are), they would be $a = 3$, $b = -1$, and $c = -6$.

Use these values to plug into the appropriate spots in the quadratic equation:

$$x = \frac{-b \pm \sqrt{b^2 - 4ac}}{2a}$$

$$= \frac{-(-1) \pm \sqrt{[-1]^2 - 4(3)(-6)}}{2(3)}$$

Simplify:

$$= \frac{1 \pm \sqrt{73}}{6}$$

So the answer could be $\dfrac{1 + \sqrt{73}}{6}$ which is approximately 1.6.

Or, it could be $\dfrac{1 - \sqrt{73}}{6}$ which is −1.25.

Now It's Your Turn

Solve the following quadratic equations using the method suggested. The answers are located in Appendix A.

11. Solve by graphing $x^2 = 2x - 6$.

12. Solve by graphing $3x^2 - x - 12 = 0$.

13. Solve by graphing $x^2 + 2x = 3$.

14. Solve by using square roots $y^2 - 4y = 15$.

15. Solve by using square roots $15y^2 = 0$.

16. Solve by using square roots $4y^2 + 10 = 11$.

17. Solve by completing the square $x^2 + 10x - 4 = 0$.

18. Solve by completing the square $x^2 = 4x$.

19. Solve by completing the square $r^2 - 14r + 2 = 0$.

20. Solve using the quadratic formula $x^2 + 10x = 24$.

21. Solve using the quadratic formula $r^2 - 12r + 6 = 0$.

22. Solve using the quadratic formula $2x^2 + 16x - 6 = 0$.

Solve the following any way you want.

23. $2x^2 + 4x = 21$

24. $x^2 + 6x = -8$

Be a Math Translator: Interpret the Discriminant

The quadratic formula contains the discriminant $b^2 - 4ac$. This is found under the radical symbol. The value of the discriminant can be used to determine the number of solutions to an equation and therefore the number of times the graph of the equation crosses the x-axis.

PSST—TRY IT THIS WAY

If $b^2 - 4ac > 0$ there are two solutions. There are also two x-intercepts.

If $b^2 - 4ac = 0$ there is one solution, and there is one x-intercept.

if $b^2 - 4ac < 0$ there are no solutions, and the graph does not cross the x-axis.

Example 11

Determine the number of solutions and x-intercepts for this equation:

$$3x^2 - 7 = 2x$$

LOOK OUT!

When asked to write an equation in standard form, rearrange so that it is in the form $ax^2 + bx + c = 0$.

Write the equation in standard form. Subtract $2x$ from each side of the equation:

$$3x^2 - 2x - 7 = 0$$

Find the value of the discriminant. In this case, $a = 3$, $b = -2$, and $c = -7$.

So ...

$$b^2 - 4ac = [-2]^2 - 4(3)\,(-7)$$
$$= 4 - -84$$
$$= 88$$

This is a positive discriminant, so the equation has two possible solutions. The graph of this equation is on the following page. Notice how there are two places where the graph crosses the x-axis.

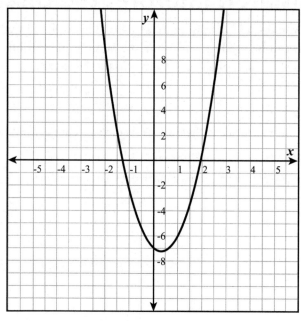

Graph of $3x^2 - 7 = 2x$.

Write the equation $y = x^2 + 3x + 10$ in standard form:

$$x^2 + 3x + 10 = 0$$

Find the value of the discriminant $b^2 - 4ac$.

In this equation, $a = 1$, $b = 3$, and $c = 10$.

So the discriminant is

$$= 3^2 - 4(1)(10)$$

$$= 9 - 40$$

$$= -31$$

The discriminant is a negative number. There is no solution, and the graph does not cross the x-axis.

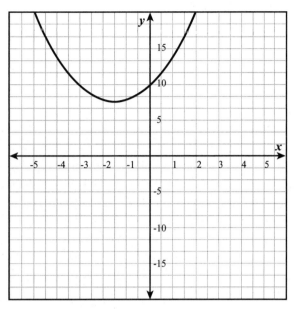

Graph of y = x² + 3x + 10.

In the equation, $8x^2 - 24x + 18 = 0$, $a = 8$, $b = -24$, and $c = 18$.

The discriminant is this:

$$b^2 - 4ac = (-24)^2 - 4(8)(18)$$

$$= 576 - 576$$

$$= 0$$

There is one solution to this equation and the graph crosses the *x*-axis in one location.

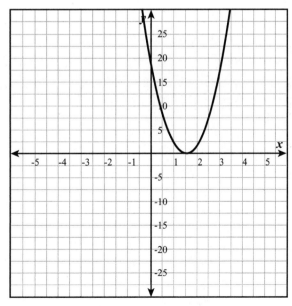

Graph of 8x² – 24x + 18 = 0.

Now It's Your Turn

For each equation, interpret the discriminant to find the number of solutions and the number of x-intercepts. The answers can be found in Appendix A.

25. $x^2 - 14x = 11$

26. $2x^2 - 8x - 7 = 0$

27. $9x^2 + 36x = -12$

28. $y = x^2 - 12x + 21$

29. $y = x^2 + 8x + 13$

30. $3x^2 + 5x = 8$

Quadratic Inequalities

In This Chapter

- Graphing quadratic inequalities with two variables
- Graphing and comparing systems of quadratic inequalities
- Solving quadratic inequalities

You have solved and graphed quadratic equalities. Now it's time to get rid of those equal signs and look at quadratic inequalities.

Graphing a Quadratic Inequality with Two Variables

There are three steps to graphing a quadratic inequality. Follow these and there will be nothing to worry about:

1. Rewrite the inequality with an equal sign. Graph the parabola that corresponds to the equation. If the original inequality is > or <, give the parabola a dotted line. If the original inequality is ≥ or ≤, give the parabola a solid line.

2. Test a point inside the parabola by plugging the values for x and y back into the original inequality.

3. If the point you tested is a solution, then shade in the portion of the graph inside the parabola. If it is not a solution, shade in the area outside the parabola.

Let's try an example.

Example 1

Graph the inequality $y > x^2 + 2x - 1$.

The first step is to turn this inequality, temporarily, into an equation. This is only for graphing purposes. And it is done by changing the > symbol into an equal sign.

You will be graphing the parabola $y = x^2 + 2x - 1$.

Notice how the original expression has a > sign. This means that when you actually draw the parabola on the graph, you should use a dotted line. Graph this equality as you have in the past.

Make a table of potential values of x and y.

x	y
1	2
0	−1
−1	−2
2	7
−2	−1

Graph the parabola for that equation.

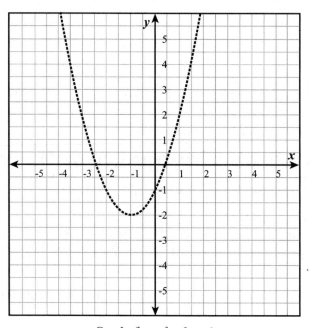

Graph of $y = x^2 + 2x − 1$.

Now, choose a point (x,y) to test. Let's do $(0,0)$ to make the math simple. Plug the value $x = 0$ and $y = 0$ into the original inequality:

$$y > x^2 + 2x − 1$$

If you do that then you are left with $0 > −1$ and that is correct. So point $(0,0)$ is a solution to the inequality. Therefore, you need to shade in the inside part of the parabola, like this:

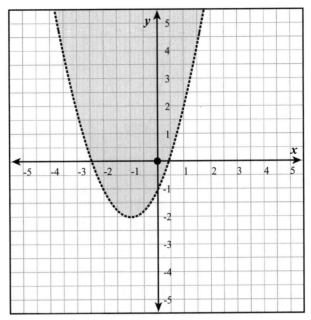

Graph of $y > x^2 + 2x - 1$ with shading.

Example 2

Graph the inequality $y \leq 2x^2 - 3x + 2$.

Start by writing this as an equation and graphing that equation:

$$y = 2x^2 - 3x + 2$$

Start a table of possible values.

x	y
1	1
0	2
−1	7
2	4
−2	16

Plot this graph.

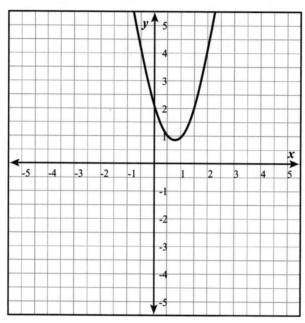

Graph of $y = 2x^2 - 3x + 2$.

Now choose a point inside the parabola to check. How about (0,3)? If that is plugged into the original inequality it will look like this:

$$y \leq 2x^2 - 3x + 2$$

$$3 \leq 2$$

Wait a second ... 3 is not less than or equal to 2. This means that the point (0,3) is not a solution to the equation. Therefore, when graphing the final inequality, the area outside the parabola should be shaded, as shown here.

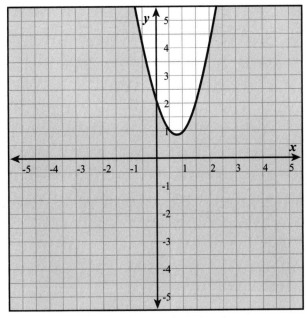

Graph of $y \leq 2x^2 - 3x + 2$.

REALITY CHECK

Quadratic inequalities may seem very esoteric and academic. However, ask any rock climber and they'll tell you, quadratic inequalities are a lifesaver (at least the ones who are fluent in algebra will!). Determining the amount of weight that a rope can safely support is an equation that requires a quadratic inequality.

Now It's Your Turn

Graph the following quadratic inequalities. You can find the answers in Appendix A.

1. $y < x^2 - 6x + 3$

3. $y \leq x^2 + 5x - 2$

2. $y \geq -x^2 + 3x - 3$

4. $y \geq x^2 + 2$

5. $y < x^2$

8. $y > 0.5x^2 - x + 2$

6. $y \geq -x^2$

9. $y + x \geq 3x^2 + 1$

7. $y < x^2 - 3x + 6$

10. $y \leq x^2 + 4x - 5$

Systems of Inequalities

You have now seen or worked through 12 different examples of how to graph inequalities. Now it is time to take it up a notch. It is time to graph systems of inequalities.

When faced with two quadratic inequalities, you need to graph both of them as you did in the previous section. The area where the shaded portions of the two graphs overlap is the answer to the system of inequalities. Take a look …

Example 3

Graph this system of inequalities:

$$y \leq x^2 + 4$$
$$y > x^2 - 3x - 1$$

Let's call the first one Inequality A and the second one Inequality B (for fun).

Graph Inequality A as you have done before.

$$y = x^2 + 4$$

x	y
0	4
1	5
−1	5
2	8
−2	8

Choose a point to test (0,5).

If $x = 0$ then the inequality says that $5 \le 4$. That's not true. This means the area outside of the parabola is shaded.

The end result is this graph:

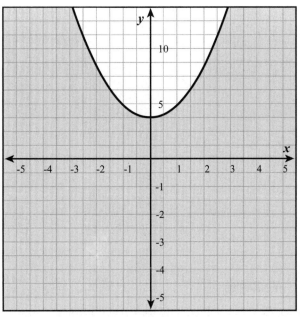

Graph of $y \le x^2 + 4$.

Use that same graph to plot the parabola for Inequality B:

$$y > x^2 - 3x - 1$$
$$y = x^2 - 3x - 1$$

x	y
0	−1
1	−3
−1	3
2	−3
−2	9

Graph this on the same graph with Inequality A. Choose a point (0,0) to see if the area inside the parabola should be shaded.

$$0 > -1$$

Yes it is! So shade in the area inside the parabola.

The graph on the next page shows what they both look like together.

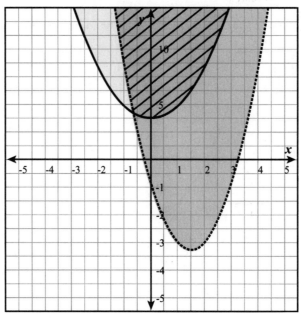

Graph of $y > x^2 - 3x - 1$.

See the portion of the graph that has the shaded areas of each inequality? This is the solution to the set of quadratic inequalities you started with.

Let's walk through another example before you are turned loose on your own!

Example 4

Graph the following system of inequalities:

$$y > x^2 - 1$$
$$y \leq -2x^2 + 4x + 3$$

Graph the first inequality:

$$y > x^2 - 1$$
$$y = x^2 - 1$$

x	y
0	−1
1	0
−1	0
2	3
−2	3

Choosing (0,0) as the point to test, you find out that $0 > -1$, which is true. Therefore, the area inside of the parabola should be shaded. This gives the graph:

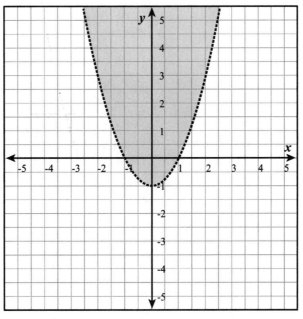

Graph of $y > x^2 - 1$.

Now add the graph for the second inequality $y \leq -2x^2 + 4x + 3$ to this one:

$$y = -2x^2 + 4x + 3$$

Make your table of possible values for x and y.

x	y
0	3
1	5
−1	−3
2	3
−2	−13

Choose point (0,0) to test and this shows that $0 < 3$, which is true. The area inside the parabola should be shaded.

This gives a final product of:

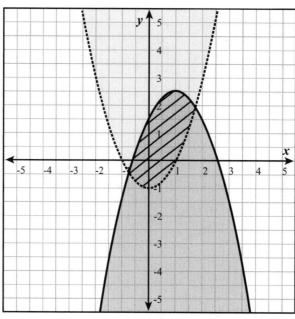

Graph of $y > x^2 - 1$ and $y \leq -2x^2 + 4x + 3$.

Now It's Your Turn

Graph each system of inequalities. The answers are located in Appendix A.

11. $y > 3x^2$ and $y \leq -x^2 + 1$

14. $y \geq x^2 - 9$ and $y \leq -3x^2 + 6x + 2$

12. $y \geq -6x^2$ and $y > 3x^2 + 3$

15. $y > 2x^2 + 2x + 4$ and $y > 3x^2 - 3$

13. $y \leq x^2 + 5x - 5$ and $y < 2x^2 + -6$

Solving Quadratic Inequalities

You may remember that there are several different ways to solve a quadratic equality. Well, there just happens to be several options for solving quadratic inequalities as well. Some of this may look familiar to you by now, but practice makes better athletes, musicians, and algebra students.

Solving a Quadratic Equation with a Table

One thing you are going to notice about the inequalities in this section is that they only have one variable. This can actually make them easier to solve.

You've made a lot of tables lately. And you are not done.

Example 5

Solve the inequality $x^2 + x \leq 6$ using a table. First rewrite the inequality so that the statement is ≤ 0. This is done by subtracting 6 from each side. Now you have this:

$$x^2 + x - 6 \leq 0$$

Make a table of values for x and for the corresponding value of $x^2 + x - 6$.

x	$x^2 + x - 6$
1	−4
−1	−6
0	−6
2	0
−2	−4
3	6
−3	0

How do you get an answer from that? Think back to the original inequality that you rearranged. It was $x^2 + x - 6 \leq 0$. This is correct when the value for x is between −3 and 2. This answer can be written as $-3 \leq x \leq 2$. There are a lot of numbers to keep track of and it is easy to get lost.

Example 6

Solve the quadratic inequality $x^2 - 2x - 8 \geq 0$ using a table.

Make a table for values of x and $x^2 - 2x - 8$.

x	$x^2 - 2x - 8$
0	-8
1	-9
-1	-5
2	-8
-2	0
3	-5
-3	7
4	0
-4	16

PSST—TRY IT THIS WAY

Finding the answer or the range of the answers for an inequality is much easier if you have two values for x that make the equation equal to 0. My advice is to keep plugging numbers in until you have two that make it equal 0. Writing the answer will be much easier!

Therefore, the values of 4 and –2 make the statement true:

$$x \leq -2 \text{ or } x \geq 4$$

LOOK OUT!

Sometimes the solution to a quadratic inequality is a range such as $-5 < x < 5$. Other times the answer will be like the answer to Example 6: $x \leq -2$ or $x \geq 4$. Keep track of the signs and make sure you are reading the table correctly.

Solving a Quadratic Inequality by Graphing

When solving $ax^2 + bx + c < 0$, set the equation equal to $y = ax^2 + bx + c$ and graph that equation first. It will be important to keep track of the signs and the symbols in this approach, too. Take a look at the next example.

Example 7

Solve $2x^2 + 2x \leq 3$ by graphing. As with the other graphing problems, rewrite the inequality in standard form:

$$2x^2 + 2x - 3 = 0$$

Find the x-intercepts by using the quadratic formula to solve for x:

$$x = \frac{-b \pm \sqrt{b^2 - 4ac}}{2a}$$

$$x = \frac{-2 + \sqrt{2^2 - 4(2)(-3)}}{2(2)}$$

$$x = \frac{-2 \pm 2\sqrt{7}}{4}$$

$x = -1.823 < x < 0.823$ or $x = 0.15$

Graph the parabola that has these intercepts.

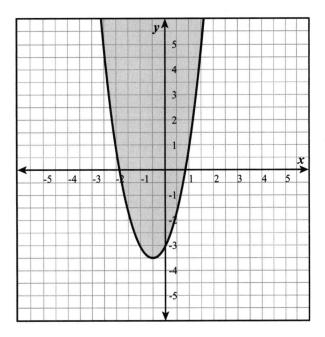

Let's try one more of those because that was so fun!

Example 8

Solve using a graph.

Solve Quadratic Equations Using Algebra

Some quadratic inequalities are most easily solved by straight algebra (these are my personal favorites). Take a look at these examples.

Example 9

Solve $x^2 - 11x > -28$ algebraically. Write the inequality as an equality by changing the > sign to an = sign:

$$x^2 - 11x = -28$$

Now write the equation in standard form:

$$x^2 - 11x + 28 = 0$$

Factor:

$$(x - 7)(x - 4) = 0$$

Solve:

$$x = 7 \text{ or } x = 4$$

These are what are called critical points to the inequality $x^2 - 11x > -28$.

Plot the numbers on a number line. Keep in mind that the inequality has a > sign. The values of 4 and 7 do not equal the answer, so keep the circles open.

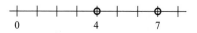

Number line with 4 and 7.

I recommend testing three points. Have one be less than 4 (try 3), one between 4 and 7 (try 5), and one greater than 7 (let's try 8). Plug each value into the original inequality to see if that particular value makes the inequality valid.

$x^2 - 11x > -28$

$x = 3$	$x = 5$	$x = 8$
$9 - 33 > -28$	$25 - 55 > -28$	$64 - 88 > -28$
$-24 > -28$ Yes!	$-30 > -28$ No!	$-24 > -28$ Yes!

So the solution to the inequality is $x < 4$ or $x > 7$.

Example 10

Using algebra, solve $x^2 - 5x > -4$.

Turn this inequality into an equation and write in standard form:

$$x^2 - 5x = -4$$
$$x^2 - 5x + 4 = 0$$

Factor:

$$(x - 1)(x - 4)$$

Solve: $x = 1$ or $x = 4$ are potential answers. Plot these on a number line.

Number line with x = 1 and x = 4.

Pick out some numbers to see what numbers are included in the solution.

Choose 0, 3, and 5 to plug into the original inequality.

$x^2 - 5x > -4$

$x = 0$	$x = 3$	$x = 5$
0 > −4 YES!	9 − 15 > −4	25 − 25 > −4
	−6 > −4 NO!	0 > −4 YES!

So the solutions for this inequality are $x < 1$ or $x > 4$.

Now It's Your Turn

Solve the following inequalities by making a table. The answers are located in Appendix A.

16. $x^2 - 7x < 12$

19. $x^2 - 3x \geq 12$

17. $2x^2 - 5x - 10 > 11$

20. $8x^2 - 3x - 9 < 0$

18. $x^2 + 2x - 3 < 0$

Solve the following inequalities by graphing.

21. $3x^2 - x - 4 \geq 0$

24. $x^2 - 5 < 0$

22. $-6x^2 - 19x \geq 10$

25. $x^2 + 6x + 3 \geq 0$

23. $x^2 - 7x + 4 > 5x + 2$

Solve the following quadratic inequalities by algebra.

26. $8x^2 - 85x + 1 < -11$

29. $-4x^2 - x + 3 \leq 0$

27. $x^2 \leq 16$

30. $2x^2 - 6x - 4 > 0$

28. $x^2 - 8x \geq -12$

Can We Please Just Be Rational Here?

Hope you liked fractions when we did them before. They are back again, and they are as ugly as ever. Rational equations and rational inequalities can be added, subtracted, multiplied, and divided. There is some flip flopping that has to occur but that can be fun, too. Just keep in mind that numerators and denominators can be pretty complex. But not so complex that you can't handle it at this point.

Fractions with Polynomials—Sounds Rational

In This Chapter

- Simplifying rational expressions
- Adding and subtracting rational expressions using the LCM
- Multiplying and dividing rational expressions
- Understanding complex fractions and simplifying them into more reasonable forms

It's time to sit back and have a rational discussion about algebra. Okay, it's not that dramatic. This chapter is an introduction to rational expressions and the manipulation of those expressions. Let's take a look.

Rational Expressions

I'd like to introduce you to one of countless different *rational expressions* you will encounter: $\frac{3}{2y+8}$.

This is a rational expression. The denominator cannot be zero. In fact, if the denominator is 0, the rational expression is undefined (not very rational at all!). Sometimes you can actually calculate the number that makes the denominator 0. That number is called the *excluded value*.

 DEFINITION

A **rational expression** is any expression that can be written as a ratio of two polynomials.

In the example given previously ($\frac{3}{2y+8}$), the excluded value is –4. Try it yourself. Plug –4 in for y in the expression. It turns the denominator into 0 doesn't it? Hmmm.

Be Rational and Keep Things Simple

It is helpful to know how to simplify rational expressions. Follow these steps to make this possible:

1. Factor the numerator.
2. Factor the denominator.
3. Divide out any common factors.

Example 1

Simplify this rational expression:

$$\frac{z}{5z}$$

In this case, you divide the common factor. The common factor here is z. Divide the z out of the numerator and the denominator and simplify. The answer is this:

$$\frac{z}{5z} = \frac{1}{5}$$

Try one more thing here. What number can z be equal to in order to make this rational expression undefined, z would have to be equal to 0. Then that means that 0 is the excluded value here. Okay, that was pretty straightforward. Let's take a look at another one.

Example 2

Simplify the rational expression:

$$\frac{8m^3 - 16m^2}{24m^2}$$

This is a bit more complicated. Factor the numerator and the denominator. And hopefully, at the end of that there will be some common factors (but there are no guarantees).

Factoring the numerator and the denominator gives the following:

$$\frac{8m^2(m - 2)}{8 \times 3 \times m^2}$$

Notice the common factors? There is an $8m^2$ in both the numerator and the denominator. Divide those out and you are left with this:

$$\frac{m - 2}{3}$$

PSST—TRY IT THIS WAY

So how do you know which factors to use? In the example here, 8 and 3 are used as factors of 24. Why not 12 and 2 or 6 and 4? Remember that you want to be able to cancel out as many factors between the numerator and denominator as possible. Pick factors that are the same whenever possible. In this case, you already had an 8 factored in the numerator. And as luck would have it, 8 is a factor of the number in the denominator, too.

Continue on here and determine what the excluded value is for the expression $\dfrac{8m^3 - 16m^2}{24m^2}$.

The only way for the denominator to be zero is if $m = 0$ (excluded values are not always 0!).

LOOK OUT!

Excluded values are not always 0. Keep that in mind. For example, in the expression $\dfrac{5}{2x+14}$ the excluded value is –7. That is the number which makes the denominator 0.

If you are asked to find the excluded value, find it using the original expression. Don't use the simplified form.

Now It's Your Turn

Simplify the following rational expressions and then give the excluded value (if there is one). The answers can be found in Appendix A.

1. $\dfrac{3x^2}{21x^4}$

2. $\dfrac{3y}{y-5}$

3. $\dfrac{x^2+4x}{3x+12}$

4. $\dfrac{9x}{9x^3+18x^2}$

5. $\dfrac{m^2-16}{m^2-8m-48}$

6. $\dfrac{y^2+4y+3}{y^2+8y+15}$

7. $\dfrac{y+2}{y^2-81}$

Adding and Subtracting Rational Expressions

Adding and subtracting rational expressions can be fairly straightforward or slightly more complicated. It all depends on what the denominators look like. Take a look at the examples in the following section.

Example 3

Find the sum of these rational expressions:

$$\frac{6}{4x} + \frac{1}{4x}$$

The denominators are the same here. Simply add the expressions, very much like you would add two fractions with the same denominator:

$$\frac{6}{4x} + \frac{1}{4x} = \frac{7}{4x}$$

There. Done. On to the next problem. Of course, they aren't all as straightforward.

Example 4

Find the differences of these rational expressions:

$$\frac{7}{2x^2} - \frac{4}{8x^3}$$

There is not a common denominator here. This means you have to move on to Plan B. You must rewrite the expression using the *least common denominator*.

DEFINITION

When dealing with rational expressions, the **least common denominator (LCD)** is the product of the factors of the denominators. It is important to use each factor only once.

When finding the LCD, one must first find the least common multiple (LCM) of the denominators. Take a closer look at the denominators of the previous rational expression. Find the factors of each:

$$2x^2 = \mathbf{2} \times x \times x$$
$$8x^3 = \mathbf{2} \times 2 \times 2 \times x \times x \times x$$

The common factors of these are bold faced. When finding the LCM, be sure to count the common factors only once. So the LCM of these two numbers are $2 \times x \times x \times 2 \times 2 \times x = 8x^3$. The LCD of $\frac{7}{2x^2}$ and $\frac{4}{8x^3}$ is $8x^3$.

Only halfway done. The question asks you to subtract those two expressions. Rewrite the fractions using the LCD of $8x^3$. In other words, change each of the fractions, multiplying by whatever it takes, to make the denominator $8x^3$. In the first fraction, in order for the denominator to be $8x^3$, the numerator and the denominator both have to be multiplied by $4x$. This is what we will do next.

The second fraction does not need anything special done to it as the denominator is already in the correct form. The new expression is now as follows:

$$\frac{7}{2x^2} - \frac{4}{8x^3} = \frac{7(4x)}{2x^2(4x)} - \frac{4}{8x^3} = \frac{28x}{8x^3} - \frac{4}{8x^3}$$

Subtract the two to get the answer:

$$\frac{28x - 4}{8x^3}$$

Whew. I think you need one more example here, don't you?

Example 5

Add these expressions:

$$\frac{3}{x+3}+\frac{x+9}{5x-4}$$

Step 1 is to find the LCM of $(x + 3)$ and $(5x - 4)$. Those two expressions cannot be factored. Therefore, they do not have any factors in common. The LCM is the product of them or $(x + 3)(5x - 4)$.

So now go back to the original addition problem:

$$\frac{3}{x+3}+\frac{x+9}{5x-4}$$

Rewrite each expression so that the denominator is in the form $(x + 3)(5x - 4)$:

$$\frac{3(5x-4)}{(x+3)(5x-4)}+\frac{(x+9)(x+3)}{(x+3)(5x-4)}$$

Simplify the numerators and the denominators:

$$\frac{15x-12}{(x+3)(5x-4)}+\frac{x^2+12x+27}{(x+3)(5x-4)}$$

Add the fractions for the answer:

$$\frac{3}{x+3}+\frac{x+9}{5x-4}=\frac{x^2+27x+15}{(x+3)(5x-4)}$$

Now It's Your Turn

Add or subtract each expression. You can find the answers in Appendix A.

8. $\dfrac{7}{3x}+\dfrac{8}{4x^2}$

9. $\dfrac{18}{x+4}+\dfrac{18}{x-4}$

10. $\dfrac{3x-5}{x^2+2x-8}-\dfrac{x+1}{x^2-4}$

11. $\dfrac{8}{3x^3}-\dfrac{5}{15x^2}$

12. $\dfrac{9}{4x}-\dfrac{1x}{x+7}$

13. $\dfrac{3}{x^2+4}+\dfrac{7x}{x+2}$

14. $\dfrac{y+4}{9y} + \dfrac{2y}{y-5}$

15. $\dfrac{4}{y^2 - 6y + 9} + \dfrac{-y}{y^2 - 11y + 24}$

Multiplying and Dividing Rational Expressions

There are two generalized equations you have to remember when multiplying or dividing rational expressions. These are …

$$\frac{a}{b} \times \frac{c}{d} = \frac{ac}{bd}$$

and

$$\frac{a}{b} \div \frac{c}{d} = \frac{a}{b} \times \frac{d}{c} = \frac{ad}{bc}$$

Try these out as you work through the next couple of examples. They really work!

Example 6

Find the product:

$$\frac{8x^2}{3x} \times \frac{3x^3}{x}$$

Multiply and you will get this:

$$\frac{(8x^2)(3x^3)}{(3x)(x)} = \frac{24x^5}{3x^2}$$

Now factor and divide out the factors that are common to the numerator and the denominator:

$$\frac{8 \times 3 \times x^5}{3x^2} = 8x^3$$

Try another example, just for fun.

Example 7

Find the product:

$$\frac{2x^2 + 4x}{3x^2 + 9x + 6} \times \frac{x^2 - 5x - 6}{x^2 - x}$$

Here is a problem with rational expressions and polynomials. The same steps apply here.

Multiply the numerators and the denominators:

$$\frac{(2x^2 + 4x)\ \ (x^2 - 5x - 6)}{(3x^2 + 9x + 6)\ (x^2 - x)}$$

Now you must factor. This looks like this:

$$\frac{2x(x+2)(x+1)(x-6)}{3x(x+2)(x+1)(x-1)}$$

Whew. There are common factors that can be cancelled out in both the numerator and the denominator. This leaves the answer to be this:

$$\frac{2(x-6)}{3(x-1)}$$

Try a little division of rational expressions.

Example 8

Find the quotient:

$$\frac{9x^2 - 9x}{x^2 - x - 6} \div \frac{x+2}{x^2 - 2x - 8}$$

First thing in a division problem is to flip it over so you have a multiplication problem (or in more acceptable algebraic language, multiply by the inverse).

Now you have the following:

$$\frac{9x^2 - 9x}{x^2 - x - 6} \times \frac{x^2 - 2x - 8}{x+2}$$

Multiply the numerators and denominators:

$$\frac{(9x^2 - 9x)(x^2 - 2x - 8)}{(x^2 - x - 6)(x+2)}$$

Factor both the numerator and the denominator:

$$\frac{9x(x-1)(x-4)(x+2)}{(x-3)(x+2)(x+2)}$$

Cancel out the common factors and the answer is this:

$$\frac{9x(x-1)(x-4)}{(x-3)(x+2)}$$

Now It's Your Turn

Find the product or the quotient of each of the following. The answers are in Appendix A.

16. $\dfrac{6x}{x^2 + 3x - 10} \times (x + 5)$

17. $\dfrac{6x}{x^2 + 2x - 8} \times (x + 4)$

18. $\dfrac{b^2 - 9b + 14}{5b} \div (b - 7)$

19. $\dfrac{2w^2 + 2w}{w^2 - 4} \div \dfrac{2w + 22}{6w - 3w^2}$

20. $\dfrac{x^2 - 9}{15x} \div (x + 3)$

21. $\dfrac{x + 2}{-x^2 - 5} \times \dfrac{x^2 + 5x}{x^2 + 4}$

22. $\dfrac{25y^2}{4} \div \dfrac{16}{5y}$

23. $\dfrac{a - 3}{a + 7} \times \dfrac{a}{2 - a}$

Complex Fractions and Rational Expressions

Moving on to complex fractions. Because keeping track of all those factors in the multiplication and division examples was not complex enough!

A complex fraction is a fraction that has another fraction in the numerator, or perhaps in the denominator, or even in both the numerator and the denominator. These beasts can look like this:

$$\dfrac{\dfrac{5}{w}}{3}$$

or like this: $\dfrac{\dfrac{x}{7}}{2}$

or even like this: $\dfrac{\dfrac{y}{4}}{\dfrac{y}{5}}$

PSST—TRY IT THIS WAY

You might want to think of simplifying a complex fraction in algebraic terms. Here is the equation:

$$\frac{\dfrac{a}{b}}{\dfrac{c}{d}} = \frac{a}{b} \div \frac{c}{d} = \frac{a}{b} \times \frac{d}{c}$$

You must divide the numerator by the denominator. And then, of course, when dividing you must multiply by the inverse.

Example 9

Simplifying these complex fractions is thankfully more simple than complex. Take a look at this example. Simplify this complex fraction:

$$\frac{\dfrac{x}{2}}{-4x^2}$$

Write the fraction as a quotient. That means that you actually divide the numerator by the denominator:

$$\frac{x}{2} \div (-4x^2)$$

Multiply by the inverse:

$$\frac{x}{2} \times \frac{1}{-4x^2}$$

LOOK OUT!

Sometimes when doing a difficult problem with many steps to keep track of, it is possible to lose sight of the simple things. Remember here that $-4x^2$ is actually a fraction. 1 is in the denominator of any whole number. So the fraction here is actually $\dfrac{-4x^2}{1}$.

Multiply the numerators and denominators to get this:

$$\frac{-x}{8x^2}$$

Simplify and the final result is this:

$$\frac{-1}{8x}$$

Try another, just to make sure you've got it before doing them on your own.

Example 10

Simplify this complex fraction:

$$\frac{\dfrac{-9y^4}{5}}{-18y^2}$$

Write the fraction as a quotient. This would be …

$$\frac{-9y^4}{5} \div (-18y^2)$$

The next step is to multiply by the inverse to turn this into a multiplication problem:

$$\frac{-9y^4}{5} \times \frac{1}{-18y^2}$$

Multiply the numerators and the denominators:

$$\frac{-9y^4}{5} \times \frac{1}{-18y^2} = \frac{-9y^4}{-90y^2}$$

Simplify to get the answer:

$$\frac{\dfrac{-9y^4}{5}}{-18y^2} = \frac{y^2}{10}$$

REALITY CHECK

Okay. Fractions with fractions? Really, how are these applicable to real life? Unfortunately, you probably won't like the answer. The practical application of complex fractions and the need to simplify them are limited to topics such as finding the voltage of DC currents in physics class. But although you might not use them in everyday life, you still have the satisfaction of being able to tackle such an ugly-looking problem.

Now It's Your Turn

Simplify each of the following complex fractions. You can find the answers in Appendix A.

24. $$\frac{\dfrac{-16b^4}{-4b^2}}{8b^3}$$

25. $$\frac{\dfrac{x^2+3x}{x+3}}{x^2-x}$$

26. $\dfrac{\dfrac{y^2 - 2y - 24}{3}}{\dfrac{y - 6}{12}}$

27. $\dfrac{\dfrac{18t^2}{-2}}{13t^4}$

28. $\dfrac{\dfrac{f^2 + 9f}{2f - 6}}{f^2 - 81}$

29. $\dfrac{\dfrac{4x}{3}}{-24x^4}$

30. $\dfrac{\dfrac{\dfrac{4x^2 + 7x - 4}{2x^2 + 5x - 3}}{10x^2 - 2x}}{2x^3 - 2x}$

Rational Equations

In This Chapter

- Solving rational equations
- Learning the difference between inverse variation and direct variation
- Graphing inverse variation

You can now simplify, add, subtract, multiply, and divide rational expressions. Time to move on to rational equations.

Be Rational, Please

A rational equation is one in which there is one, or possibly more, rational expressions. Many rational expressions can be solved using the *cross products property*. Take a look at the example to see how this works.

> **DEFINITION**
>
> The **cross products property** is sometimes also called cross-multiplying. It is the product of the numerator of one ratio and the denominator of the other ratio.
>
> For example, to solve for x in this proportion, you would use the cross products property: $\frac{2}{x} = \frac{8}{16}$. Cross-multiply to get $2(16) = 8x$.
>
> Solve for x and find that $x = 4$.

Example 1

Solve this rational equation:

$$\frac{2}{x-3} = \frac{x}{9}$$

Use the cross products property and multiply the numerator of one rational expression by the denominator of the other. If you do that, then you have this:

$2(9) = x(x - 3)$

$18 = x^2 - 3x$

Subtract 18 from both sides:

$$0 = x^2 - 3x - 18$$

Factor that polynomial:

$$0 = (x - 6)(x + 3)$$

You can use the zero product property to solve for x:

$$x - 6 = 0 \text{ or } x + 3 = 0$$

Solve for x:

$$x = 6 \text{ or } x = -3$$

Check your work by plugging those values of x back into the original rational equation:

$$\frac{2}{x - 3} = \frac{x}{9}$$

If $x = 6$ If $x = -3$

$$\frac{2}{6 - 3} = \frac{6}{9} \qquad\qquad \frac{2}{-3 - 3} = \frac{-3}{9}$$

$$\frac{2}{3} = \frac{2}{3} \qquad\qquad -\frac{1}{3} = -\frac{1}{3}$$

Sometimes it makes sense to use the least common denominator to solve a rational equation. Here is an example of how this is done.

Example 2

Solve the following rational equation:

$$\frac{4}{x - 8} + 1 = \frac{5}{x - 8}$$

Factor each of the denominators. The least common denominator, LCD, is $x - 8$.

Now multiply each side by the LCD:

$$\frac{4}{x - 8}(x - 8) + 1(x - 8) = \frac{5}{(x - 8)}(x - 8)$$

Multiply and simplify:

$$4 + (x - 8) = 5$$

Combine and solve for x:

$$x = 9$$

Example 3

Factor to find the LCD and then solve:

$$\frac{k}{k-2} + \frac{1}{5} = \frac{2}{k-2}$$

The LCD in this case is $5(k-2)$. Multiply each part of the problem by the LCD.

$$\frac{k}{k-2}[5(k-2)] + \frac{1}{5}[5(k-2)] = \frac{2}{k-2}[5(k-2)]$$

Simplify:

$$5k + k - 2 = 10$$

$$6k - 2 = 10$$

$$6k = 12$$

$$k = 2$$

Hold on a sec. If you plug the answer $k = 2$ into the original equation, then the denominator will be 0. 2 is an extraneous solution. There is no solution to this problem!

Now It's Your Turn

Solve each rational equation using either the cross products property or the LCD. You can find the answers in Appendix A.

1. $\dfrac{x+1}{2x+2} = \dfrac{2}{3x}$

2. $\dfrac{4}{x^2+4} = \dfrac{2-3x}{x+2}$

3. $\dfrac{5y+1}{15y-1} = \dfrac{3}{5}$

4. $\dfrac{x+1}{2x+2} = \dfrac{3}{2}$

5. $\dfrac{k}{k-6} = \dfrac{24}{k}$

6. $\dfrac{f+2}{f^2+6f-7} = \dfrac{8}{f^2+3f-4}$

7. $\dfrac{1}{m+3} + 2 = \dfrac{m^2 - 3}{m^2 + 8m + 15}$

9. $\dfrac{4}{x-3} + 1 = \dfrac{10}{x^2 + x - 12}$

8. $\dfrac{24}{x-3} = \dfrac{x}{3}$

10. $\dfrac{9}{y+1} = \dfrac{4}{y+4}$

Variations—The Big Change Up

There are a few different types of variation. The equation $y = ax$ (when $a \neq 0$) is an example of direct variation. Think of it this way: y varies directly with x. If the value of x increases, the value of y increases as well.

Inverse variation is the case when $y = \dfrac{a}{x}$ when $a \neq 0$. In this case, a is the constant of variation. Think of it like this—y varies inversely with x. So if x increases, y decreases. Let's look at one.

Example 4

Determine whether this equation is a direct variation, inverse variation, or neither:

$xy = -5$

Remember that you had two different formulas to use here. The formula for direct variation which is $y = ax$ and the formula for indirect variation which is $y = \dfrac{a}{x}$.

Rearrange the equation here so that y is isolated and let's see what you have:

$xy = -5$

Divide both sides by x:

$\dfrac{xy}{x} = \dfrac{-5}{x}$

$y = \dfrac{-5}{x}$

This is an example of inverse variation. The constant of variation is –5.

PSST—TRY IT THIS WAY

Sometimes an equation does not show either direct variation or indirect variation. How will you know when that is the case? That is the case when the equation cannot be rearranged to be written in the form $y = ax$ or $y = \dfrac{a}{x}$. For example, the equation $y = 7x - 9$ cannot be written in either of those forms. This equation does not represent direct or inverse variation.

Example 5

Determine if this equation shows direct variation, indirect variation, or neither:

$$y = 4x + 7$$

You want to rewrite this in the form of $y = ax$ or $y = \dfrac{a}{x}$ if it is to be some type of variation. That can't be done here. Because + 7 is at the end of the equation, the equation cannot be rearranged. This equation does not show any variation. And that is okay.

Try a few quick examples to make sure you know how to do it.

Now It's Your Turn

Rearrange each equation as needed. Tell if the equation represents direct variation, indirect variation, or neither. You can find the answers in Appendix A.

11. $y = 8x$

14. $4 - y = 6$

12. $x = 12y$

15. $\dfrac{y}{x} = 3$

13. $y = \dfrac{x}{2}$

Graphing an Inverse Variation Equation

Again, this process is going to be similar to what you have done for other graphs that you have drawn. Make a chart of values for x and y and then plot the points on a graph. Be sure to include positive and negative values in your chart.

Example 6

Graph this inverse variation equation:

$$y = \frac{3}{x}$$

x	y
1	3
–1	–3
2	1.5
–2	–1.5
3	1
–3	–1

Plot these points on a graph.

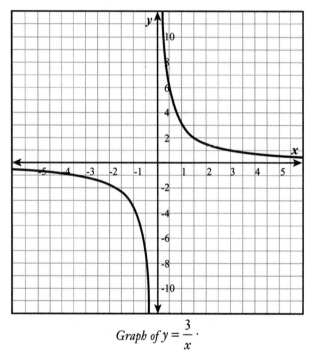

Graph of $y = \dfrac{3}{x}$.

Notice how the graph has a different look from ones that you have seen in the past. The graph moves along the *x*-axis without actually touching it. The same is true for the *y*-axis.

That shape is a hyperbola. When an inverse variation is plotted, the resulting shape is two symmetrical parts which are called the branches of the hyperbola. The lines that they come close to touching but don't ever actually do touch are called asymptotes.

Try another example.

Example 7

Graph $y = \dfrac{-2}{x}$.

Start with your table of positive and negative numbers.

x	y
1	-2
-1	2
2	-1
-2	1
4	$-\dfrac{1}{2}$
-4	$\dfrac{1}{2}$

Now graph those points.

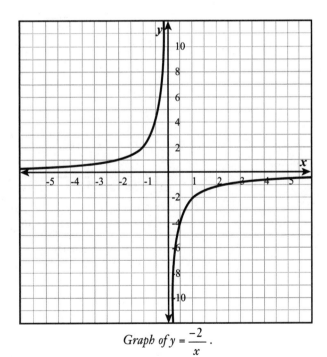

Graph of $y = \dfrac{-2}{x}$.

Hmmm. Notice anything about this graph compared to the one in the previous example? When an inverse equation has a value of $a > 0$ then the graph is in Quadrants I and III. When the inverse equation has a value of $a < 0$, the graph is in Quadrants II and IV.

LOOK OUT!

Notice how there is no value for $x = 0$ in either of the tables used for these two examples? That is because the value of y when $x = 0$ is undefined. There is no point $(0,y)$ on the graphs. The graphs get close to $x = 0$ but never actually make it there!

PSST—TRY IT THIS WAY

If you really want to get a handle on what happens to the graph as the line nears the x-axis, then plot higher or lower numbers in your table. Choose numbers very close to 0 and numbers very far from $x = 0$ to see how the graph behaves.

Now It's Your Turn

Graph the following equations of inverse variation. The answers are located in Appendix A.

16. $y = \dfrac{5}{x}$

19. $y = \dfrac{1}{x}$

17. $y = -\dfrac{7}{x}$

20. $y = -\dfrac{1}{x}$

18. $2xy = 12$

Using Inverse Variation Equations

That is always the big question in algebra isn't it? When will I ever use this information? Here are some ways in which you might use inverse variations (or at least some problems for you to solve now).

Example 8

Suppose you have two variables, x and y. These two variables vary inversely. So when $y = 4$, $x = -2$. What inverse relation could be written to describe the relationship between these two variables?

You know that they show inverse variation so their equation must be in this form:

$$y = \frac{a}{x}$$

You know that when $y = 4$, then $x = -2$. You can plug that information into the formula to solve for a (the constant of variation):

$$4 = \frac{a}{-2}$$

Solve for a:

$$a = -8$$

So the equation that relates to the inverse variation when $y = 4$ and $x = -2$ is ...

$$y = \frac{-8}{x}$$

Take this one step further and solve for y when $x = 2$:

$$y = \frac{-8}{2} = -4$$

Example 9

Here are several ordered pairs:

$$(-6,2)\ (-4,3)\ (9,-1.33)\ (24,-0.5)$$

Do these numbers show inverse variation? If so, write the equation describing the inverse variation. If not, move on to the next problem.

Look at the equation for inverse variation again:

$$y = \frac{a}{x}$$

If you were to multiply each side by x then $yx = a$. Which means that the product of an x-value and a y-value would be equal to the constant of variation.

Let's see if that is the case with these numbers:

$$-6 \times 2 = -12$$

$$-4 \times 3 = -12$$

$$9 \times -1.33 = -12$$

$$24 \times -0.5 = -12$$

All of the ordered pairs, when multiplied, are equal to –12. This means that $a = -12$.

Rewriting the equation for inverse variation with this constant would give us the following:

$$y = \frac{-12}{x}$$

Example 10

Here is a set of ordered pairs:

$(-2,5)$, $(2.5,-4)$, $(1.25,-8)$, $(-0.5,20)$

Do these numbers represent inverse variation? If so, write the equation for the inverse variation.

Keep in mind the formula for inverse variation: $y = \frac{a}{x}$ that can be rewritten as $yx = a$. If these ordered pairs show inverse variation, their product will all be the same:

$-2 \times 5 = -10$

$2.5 \times -4 = -10$

$1.25 \times -8 = -10$

$-0.5 \times 20 = -10$

They all equal –10 so it is an inverse variation. The equation is $y = \frac{-10}{x}$.

Now It's Your Turn

Graph each inverse equation. You can find the answers in Appendix A.

21. $xy = -1$

23. $2x = \frac{10}{y}$

22. $y = \frac{-5}{x}$

Suppose that y varies inversely with x. Write the inverse variation equation that relates the two.

24. $x = 8, y = 2$

26. $x = 6, y = 12$

25. $x = -2, y = -12$

Tell if each set of ordered pairs represents inverse variation. If so, write the formula.

27. (4,1), (8,2), (12,3), (16,4), (30,4)

29. (−10,−30), (−5,−60), (40,7.5), (20,15)

28. (−24,1), (−10,2.4), (5,−4.8), (−4,6)

Odds and Ends, Data and Words

Many of you will find this last part to be the most applicable to real life. You can think of this as the point when you take what you have worked so hard on and actually *use* it. I am not making promises that this part will help you double your money at the casino or help you land a better interest rate on your mortgage. But the problems you solve here may help you understand why the odds are against you at the casino, why it seems as if your savings account is earning money at a snail's pace, and why it always seems like your uncle gets to your family dinner way before your cousins do.

Probability and Odds

In This Chapter

- Solving problems involving theoretical and experimental probability
- Determining the odds for or against a particular event
- Applying the counting principle to permutation problems
- Learning the difference between independent and dependent compound events

The past few chapters were pretty intensive with all the graphing and formulas. Time for a little fun. This chapter looks at probability and odds. It's not meant to make your money work for you in Vegas but what you learn here could help you understand why a 65 percent chance of rain doesn't always mean you are going to need your umbrella. Or why forgetting your ATM code can be disastrous. Or why packing for a vacation can be more time-consuming than you planned.

The probability of something occurring is a measurement of the likelihood, or the chance that a certain event might occur. Probability is a real number between 0 and 1. A probability P of 0 means it is impossible that something will happen. A probability of 0.5 means it is equally likely that something will or will not happen. P = 1 means a certain event is going to happen.

Theoretical and Experimental Probability and Odds

There are three different types of probability. A summary of each is provided here.

Theoretical probability is defined as $\dfrac{\text{number of favorable outcomes}}{\text{total number of outcomes}}$.

LOOK OUT!

Probability can be expressed as a decimal, as a fraction, or as a percent. Don't be surprised if you see it expressed in any of these ways.

Example 1

Suppose the letters I L O V E M A T H were each written on a small piece of paper. The slips of paper are placed in a hat and one letter is selected at random. What is the theoretical probability of selecting a piece of paper with the letter A on it?

First off, you know the chance of selecting the card with the letter A is equally as likely as choosing any of the other cards.

Remember the formula:

$$\text{theoretical probability} = \frac{\text{number of favorable outcomes}}{\text{total number of outcomes}}$$

Look at the letters. How many possible outcomes are there? There are 9 different letters, so there are 9 possible outcomes. That will be put in the denominator of the equation. How many outcomes are there for drawing the letter A? Only one. There is only one A in the whole batch of numbers.

So the theoretical probability that you will draw the letter A from the hat is $\frac{1}{9}$.

PSST—TRY IT THIS WAY

There are several different ways you can talk about this solution. You can say "there is a 1 in 9 chance that I draw the letter A." You can say "there is a 0.111 chance that I draw the letter A." Or you can say, "I have an 11 percent chance of drawing the letter A."

Example 2

What is the theoretical probability of rolling a number greater than or equal to 5 on a standard die (or number cube)? There are six sides to your standard die. These are numbered 1–6 so there are 2 possible sides that meet the requirements of the question.

That means:

$$\text{theoretical probability} = \frac{\text{number of favorable outcomes}}{\text{total number of outcomes}} = \frac{2}{6}$$

So if you roll the die 6 times, you are likely to get a number ≥ 5 twice. Or if you roll it 3 times, then you are likely to get a number ≥ 5 once.

Experimental probability is expressed as

$$P(\text{some event}) = \frac{\text{number of successes}}{\text{number of trials}}.$$

It is important to keep in mind what you are measuring. The number of favorable outcomes is the number of successes.

Example 3

Tristan flips a penny 40 times. He gets tails 15 times. What is the experimental probability that the penny will come up tails?

Use the following formula:

$$P(\text{some event}) = \frac{\text{number of successes}}{\text{number of trials}}$$

The successes in this example are the number of times that the penny comes up tails. That is the favorable outcome. Tristan's probability is equal to $\frac{15}{40}$ because his coin came up tails 15 times out of 40. The fraction can be simplified:

$$P(\text{tails}) = \frac{15}{40} = \frac{3}{8}$$

So again, Tristan can say that he got tails 3 out of every 8 times. He can say that 37.5 percent of the time he got tails. Or he could say that the chances of his coin landing on tails is 0.375.

Example 4

An overzealous algebra student tosses a number cube 500 times. Seventy-eight times, the cube lands on the number 3. What is the experimental probability that the cube lands on 3?

The formula you use here is

$$P(\#3) = \frac{\text{number of successes}}{\text{number of trials}} = \frac{78}{500} = \frac{39}{250} = 0.156$$

This is about 16 percent. So the number 3 will come up roughly 16 percent of the time.

Are the Odds with You?

Sometimes you want to compare the favorable and the unfavorable outcomes. If that is the case, then you are looking at the odds of an event.

The odds in favor of a particular outcome are …

$$\frac{\text{number of favorable outcomes}}{\text{number of unfavorable outcomes}}$$

The odds against an outcome are …

$$\frac{\text{number of unfavorable outcomes}}{\text{number of favorable outcomes}}$$

Example 5

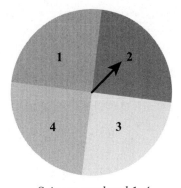

Spinner numbered 1–4.

What are the odds against stopping on "1"? Look carefully at what you are trying to determine. Here you want to find the odds against something.

Then you want to use this equation: $\dfrac{\text{number of unfavorable outcomes}}{\text{number of favorable outcomes}}$.

In this scenario, you are equally as likely to stop on 1, 2, 3, or 4. Landing on "1" is favorable, so the other three options are unfavorable.

Therefore, the odds are equal to $\dfrac{3}{1}$ or 3:1.

> **LOOK OUT!**
>
> Odds are typically expressed as a ratio. For example, the odds of $\dfrac{3}{1}$ are 3 to 1.

Example 6

A spinner has six different colors on it—red, orange, blue, purple, green, and yellow. Each color covers the same amount of the face of the spinner. What are the odds that you do not land on red or orange?

You look for the odds that something *doesn't* happen here. So the equation to use is …

$$\dfrac{\text{number of unfavorable outcomes}}{\text{number of favorable outcomes}}$$

There are two potential unfavorable outcomes, landing on red or landing on orange. That leaves four favorable outcomes (landing on blue, purple, green, or yellow). The formula is:

$$\dfrac{\text{number of unfavorable outcomes}}{\text{number of favorable outcomes}} = \dfrac{2}{4} = \dfrac{1}{2}.$$

Now It's Your Turn

Determine the theoretical probability, the experimental probability, or the odds of each situation given. The answers are located in Appendix A.

1. The Tasty Cat Food Company has had trouble lately with their Tuna Surprise. The board of health checked 400 cans and found that 18 of the cans contained dangerous levels of bacteria. What is the experimental probability that any can of Tuna Surprise is dangerous?

2. What is the probability of rolling an odd number using a standard die?

3. Becca spins a spinner with the numbers 1 to 9 on it. What is the probability of it stopping on an odd number?

4. There are three children in the Green family. What is the probability that all three of the children are boys?

5. Zoe has a spinner with 26 divisions— one for each letter of the alphabet. What are the odds that the spinner will land on the letter Z when she spins it?

6. The local youth baseball team is having a raffle to win tickets to a Red Sox game. One ticket out of 350 will be drawn on the day of the tournament. If you bought one ticket, what are the odds that you will win? What is the probability that you will win?

7. Suppose the odds in favor of some event are $\frac{7}{9}$. What are the odds against it?

8. You roll a die. What are the odds of rolling a number greater or equal to 3?

9. A spinner has 10 different divisions numbered 1–10. Bill spins it 25 times. It stops on the number 7 four times. What is the experimental probability of stopping on number 7?

10. The local meteorologist said there was a 60 percent chance of snow this afternoon. What are the odds that it is going to snow?

Permutations

Permutations are arrangements of objects where the order in which they are arranged is important. Suppose you were given the letters C-A-T. How many different ways can you arrange those letters? Here they are:

C-A-T, C-T-A, A-C-T, A-T-C, T-A-C, T-C-A

There are six unique ways these letters can be arranged.

Example 7

How many different ways can the letters in the word M-A-T-H be arranged (how many permutations are there)? Use the *counting principle* to determine how many permutations are in the letters M-A-T-H.

> **DEFINITION**
>
> The **counting principle** says that one event can occur in m different ways. For each of these ways, a second event can occur in n ways. The number of ways that the two events can occur together is $m \times n$.
>
> If a deli has two possible soups for lunch and four possible sandwiches, then a lunch special of soup and a sandwich has a total of eight different choices. Two soup choices × four sandwich choices is equal to eight different options.

In this example, the number of permutations is equal to the possible letters for the first letter × the choices for the second letter × the choices for the third letter × the choices for the fourth letter.

Number of permutations = 4 × 3 × 2 × 1 = 24

There are 24 possible combinations. For those of you who are doubtful, they are listed here:

MATH	ATHM	TAHM	HATM
MAHT	ATMH	TAMH	HAMT
MTAH	AHTM	THAM	HMAT
MTHA	AHMT	THMA	HMTA
MHAT	AMTH	TMAH	HTAM
MHTA	AMHT	TMHA	HTMA

That answer, 4 × 3 × 2 × 1 is called a factorial. In fact it is the 4 factorial. Mathematicians get very excited when they talk about factorials and they tend to speak very loudly about them. Factorials can be written in the general format $n!$. This is the product of 1 to n.

In other words $n! = n \times (n-1) \times (n-2) \times \ldots \times 3 \times 2 \times 1$

Another useful way of looking at this is to solve for the number of permutations of n objects taken r at a time.

$$_nP_n = \frac{n!}{(n-r)!}$$

> **PSST—TRY IT THIS WAY**
>
> The value of $0! = 1$.
>
> The number of permutation for n object is equal to
>
> $_nP_n = n!$
>
> So the number of permutations of 4 objects (as in the previous example) is $_4P_4 = 4! = 4 \times 3 \times 2 \times 1 = 24$.

Example 8

Anne has 15 different T-shirts. She needs to bring 9 of them with her on vacation. How many possible combinations of T-shirts could she bring with her? You are trying to find the number of permutations of 9 T-shirts chosen from a total of 15. In other words, you want to find $_{15}P_9$.

Use this formula:

$$_nP_n = \frac{n!}{(n-r)!}$$

$$= \frac{15!}{(15-9)!}$$

Subtract the numbers in the denominator:

$$= \frac{15!}{6!}$$

Expand the factorial until you reach a common factorial:

$$= \frac{15 \times 14 \times 13 \times 12 \times 11 \times 10 \times 9 \times 8 \times 7 \times 6!}{6!}$$

Cancel out the common factor (6!) and solve:

$$= 1{,}816{,}214{,}400$$

There are 1,816,214,400 different T-shirt combinations when trying to take 9 out of 15.

Now It's Your Turn

Solve each permutation problem. The answers are located in Appendix A.

11. How many different ways can you arrange the letters in HORSE?

12. How many different ways can you arrange the letters in BACTERIUM?

13. Evaluate $_8P_8$.

14. Evaluate 0!.

15. Evaluate $_{15}P_0$.

16. The cheerleaders each make a poster with one of the following letters on it: C-O-U-G-A-R-S. If the posters are distributed randomly, what is the probability that the cheerleaders will correctly spell COUGARS for their cheer?

17. Alina has the following ingredients for a fruit salad at home: strawberries, blueberries, raspberries, apples, grapes, mangoes, and bananas. If she wants to use three different ingredients in a fruit salad, how many different possible combinations are there?

18. LeAnn is making a piece of stained glass to hang in her window. The pattern she is using has six different spots. She has eight different colored glasses she could use. How many different color combinations are there?

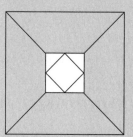

19. Fiona has a 4-digit pin number for her ATM card. She knows that the digits are four different numbers from 5–9. She forgot the exact code though and has to guess. What is the probability that she will guess correctly?

Compound Events

Sometimes you hear about events that are a little different. You might order off a Chinese takeout menu and have the choice of ordering the lo mein and the egg roll or the lo mein and the fried rice.

These are *compound events*. Compound events can be mutually exclusive and represented by the equation P(A or B) = P(A) + P(B).

Or they can be overlapping events and be represented by the formula:

$$P(A \text{ or } B) = P(A) + P(B) - P(A \text{ and } B)$$

> **DEFINITION**
>
> A **compound event** combines two or more events using the words "and" or "or." Mutually exclusive events have no outcomes in common. Overlapping events have at least one common outcome.

Example 9

Lilia rolls a die. What is the probability that she will roll an even number or a prime number?

Step back a second and think about this. There are three even numbers on a die. There are three prime numbers. And there is one number that overlaps (2 is a possible answer for both because it is even and prime). Because these are overlapping events, you should use this formula:

$$P(A \text{ or } B) = P(A) + P(B) - P(A \text{ and } B)$$

$$P(\text{prime or even}) = P(\text{even}) + P(\text{prime}) - P(\text{prime and even})$$

$$= \frac{3}{6} + \frac{3}{6} - \frac{1}{6}$$

$$= \frac{5}{6}$$

Some compound events are *independent events* and others are *dependent events*.

> **DEFINITION**
>
> **Independent compound events** occur if one event has no impact on another. You have a bag of red and green marbles. You randomly choose a red marble and then put the marble back into the bag before choosing again. These are independent events.
>
> **Dependent compound events** have an impact on each other. You have the same bag of marbles. But this time after drawing the red marble you put that on the table before reaching into the bag again. The bag of marbles has changed and these are dependent events.

Independent events are calculated using this formula:

$$P(A \text{ and } B) = P(A) \times P(B)$$

Dependent events are calculated using this formula:

$$P(A \text{ and } B) = P(A) \times P(B \text{ given } A)$$

We'll try one more example and then you are on your own!

Now It's Your Turn

Answer each of the following questions. The answers are located in Appendix A.

20. A pencil box contains 5 green pencils, 8 yellow pencils, and 3 blue pencils. What is the probability that you will draw a yellow or a blue pencil?

Roll a number cube. And then find P(A or B) for the following:

21. Event A: Roll a number less than 5

    Event B: Roll an odd number

24. There are 8 red apples and 3 green apples in a bowl. You close your eyes and choose one apple. You place that apple on the counter and then randomly choose another. What is the approximate probability that both apples you chose were red?

    Identify each set of events as dependent or independent. Then solve for P(A and B).

22. Event A: Roll a 3

    Event B: Roll an even number

23. Event A: Roll a multiple of 2

    Event B: Roll an odd number

25. You roll two dice:

    Event A: You roll a 5 first.

    Event B: You roll a 1 next.

26. You flip two coins:

    Event A: Both coins are heads.

    Event B: One coin is heads and one is tails.

27. You write each letter in the word ALGEBRA on a separate slip of paper. You place the slips into a hat and randomly choose one letter. Do not replace that letter and then choose another.

    Event A: The first letter is G.

    Event B: The second letter is A.

28. Maya's closet has 4 blue skirts, 7 black skirts, 2 red skirts, and 3 green skirts. She randomly chooses a skirt, sets it aside, and then randomly chooses a different color. What is the probability that both skirts she chooses are green?

29. What if Maya first chooses a black skirt, replaced it, and then chose another black skirt the second time? What is the probability of that happening?

Organizing and Interpreting Data

In This Chapter

- Learning the measures of central tendency and dispersion
- Creating and interpreting a stem-and-leaf plot
- Creating and interpreting a histogram
- Creating and interpreting a box-and-whisker plot

This chapter is going to be a little different from some of the others in this book. Instead of formulas and fractions and calculations, this chapter will show you ways to display and organize data. You'll be asked to interpret and create graphs. This might be a nice change of pace.

Central Tendency

Ever notice how in some families, every Sunday afternoon in the fall, there is a cluster of people in the living room gathered around the television watching football (it can't be just my family where this occurs). This phenomenon has an algebraic counterpart. That is the algebraic measurement of central tendency. In algebra, central tendency is a way to describe how data clusters around one value (that value being the television).

There are three different ways that central tendency can be measured:

- Mean
- Median
- Mode

The mean of a set of numbers is the average. This is the answer you get when you add up all of something and divide by the total number of that something. Mean is expressed as $\bar{x}$ (x-bar) and is defined by this formula:

$$\bar{x} = \frac{x_1 + x_2 + x_3 + \ldots x_n}{n}$$

The median is the number value in a set of numbers. If there are an even number of values in a set, then it is the average of the two middle values.

The mode is the value that occurs most often in a set of data.

LOOK OUT!

There can be more than one mode in any given set of data. There could be only one mode. Or, quite possibly, there could be none.

Let's see how these values of central tendency can be used and what they show us.

Example 1

Mr. Michael, the physics teacher, gave an exam on Force, Mass, and Momentum last week. The following are the scores for his third period class: 88, 92, 98, 45, 56, 74, 69, 73, 64, 88, 83, 90, 81.

Find the mean, median, and mode of those test grades.

The mean is calculated using the formula:

$$\bar{x} = \frac{x_1 + x_2 + x_3 + \dots x_n}{n}$$

In this case, there are 13 different scores, so $n = 13$.

Solve for $\bar{x}$:

$$\bar{x} = \frac{88 + 92 + 98 + 45 + 56 + 74 + 69 + 73 + 64 + 88 + 83 + 90 + 81}{13}$$

$$= 77$$

To find the median, arrange the test scores in numerical order:

45 56 64 69 73 74 81 83 88 88 90 92 98

There are an odd number of scores so it is pretty easy to find the middle value of that set:

45 56 64 69 73 74 **81** 83 88 88 90 92 98

81 is the median.

The mode is 88. Two of the students scored 88 on the test. No other test score was repeated.

PSST—TRY IT THIS WAY

The median best represented the data. This is the number that is right in the middle of the whole data set. The mean in this case is less than the median and the mode is higher.

Example 2

Sometimes, teachers might drop the highest and lowest test scores to see what this does for the class average. If Mr. Michael did that for his third period class, what would happen to the mean, median, and mode?

Here are the test scores in order:

45 56 64 69 73 74 **81** 83 88 88 90 92 98

If Mr. Michael drops out the 45 and the 98, he would be left with 11 scores.

The mean of the test scores would now be this:

$$\bar{x} = \frac{56 + 64 + 69 + 73 + 74 + 81 + 83 + 88 + 88 + 90 + 92}{11}$$

$$= 78$$

The mean increases by 1 point.

The median is 56 64 69 73 74 **81** 83 88 88 90 92.

The median stays the same!

The mode is still 88. So the mean increases, but the median and mode remain the same.

Now It's Your Turn

Find the mean, median, and mode of each set of numbers. The answers are located in Appendix A.

1. 0, 1, 1, 3, 4, 6, 0, 19, 12

2. 9, 19, 29, 39, 49, 59

3. 0.7, 0.9, 0.25, 0.01, 0.15, 0.025, 1.0

4. 85, 76, 99, 65, 75, 78, 80

5. 1, 1, 1, 1, 13, 13, 15, 19

6. Jennifer wants to end the semester with an average of 89 in algebra. On her first five tests she received the following scores:

   72, 75, 92, 94, 88

   What does she have to get on the sixth and final test to meet her goal?

Dispersion

Algebraically speaking, dispersion is how much a set of data is spread out. This is discussed using two different measurements: range and mean absolute deviation. The range of a set of numbers is the difference between the highest value in the set and the lowest value in the set.

The mean absolute deviation is the average variation of the data from the mean. Mean absolute deviation is found using this equation:

$$\text{mean absolute deviation} = \frac{\left|x_1 - \bar{x}\right| + \left|x_2 - \bar{x}\right| + \left|x_3 - \bar{x}\right| + ... + \left|x_3 - \bar{x}\right|}{n}$$

Working through an example will help.

Example 3

Mrs. Berry measures the height of her preschool class. Her 10 students have the following heights, in inches:

30	36	30	42	39	35	32	33	35	35

What is the range of the heights of these students?

The tallest child is 42 inches tall. The shortest kid is 29 inches. The difference between these numbers will give the range:

$$42 - 29 = 13 \text{ inches}$$

What is the mean absolute deviation? First find the mean (or average) of the heights:

$$\bar{x} = \frac{30 + 36 + 30 + 42 + 39 + 35 + 32 + 33 + 35 + 35}{10}$$

$$= 34.7$$

Use that number to calculate the mean absolute deviation:

$$\frac{\left|30 - 34.7\right| + \left|36 - 34.7\right| + \left|30 - 34.7\right| + \left|42 - 34.7\right| + \left|39 - 34.7\right| + \left|35 - 34.7\right| + \left|32 - 34.7\right| + \left|33 - 34.7\right| + \left|35 - 34.7\right| + \left|35 - 34.7\right|}{10}$$

This is a tad cumbersome, but remember you are trying to find out how much each value deviates or varies from the average:

$$\frac{\left|-4.7\right| + \left|1 3\right| + \left|-4.7\right| + \left|7.3\right| + \left|4.3\right| + \left|0.3\right| + \left|-2.7\right| + \left|-1.7\right| + \left|0.3\right| + \left|0.3\right|}{10}$$

$$= \frac{27.6}{10}$$

$$= 2.76$$

LOOK OUT!

Remember that the absolute value of a negative number is a positive number!

$$\left|-a\right| = a$$

Now It's Your Turn

Find the range and the absolute deviation of each set of numbers. The answers are located in Appendix A.

7. 46.38, 53.69, 47.25, 49.19

10. 30, 30, 20, 25, 28, 32, 33, 29

8. 111, 113, 115, 125, 130, 145

11. 12, 15, 11, 12, 10, 9, 18, 15

9. 506, 500, 508, 510, 522, 501, 498

12. 7, 4, 6, 6, 5, 8, 3, 9

Stem-and-Leaf Plots

Stem-and-leaf plots organize data based on their digits. The stem is the leading digits and the leaf is the last digit. These are used to show how data can be distributed.

Example 4

One winter, a particular city had several significant snow falls. The snow totals, in centimeters, are listed here:

46   22   31   36   61   25   46   31   38   25   12

Make a stem-and-leaf plot of this data and interpret what it shows. First, choose the stems for the stem-and-leaf plot. With this particular data set, the stems will be the number in the tens place.

There are five different stems here—1, 2, 3, 4, and 6. Start the stem-and-leaf plot with that information.

Next, add the data for the leaves, which, in this case, is in the ones place.

```
1|
2|
3|
4|
6|
```

Stem-and-leaf.

The number 46 has 4 as the stem and 6 as the leaf. So add the 6 in the same row as the 4. The next number is 22. The stem is 2 and the leaf is 2. Place the 2 in the same row as the 2 stem.

Work through each number and the stem-and-leaf plot will look like this.

```
1 | 2
2 | 2 5 5
3 | 1 1 6 8
4 | 6 6
5 |
6 | 1
```

Example 5

Make a stem-and-leaf plot of the following data: 84, 127, 84, 102, 107, 85, 88, 125, 126, 119, 118, 102, 101, 91, 90, 88, 89, 103, 117, 124.

The numbers here vary from 84 to 126. The stems of the stem-and-leaf plot should be 8, 9, 10, 11, and 12. If that is the case, then add the other values in to get the stem-and-leaf plot that follows:

```
 8 | 4 4 5 8 8 9
 9 | 0 1
10 | 1 2 2 3 7
11 | 7 8 9
12 | 4 5 6 7
```

Key: 11 | 7 represents 117

Now It's Your Turn

Plot the data given on stem-and-leaf plots. You can find the answers in Appendix A.

13. 119, 128, 111, 103, 90, 93, 94, 98, 100, 102, 101, 121

14. The average life expectancy for some third world countries in 1980 were 45, 36, 53, 53, 66, 62, 72. Plot this on a stem-and-leaf plot.

15. The average length of sunfish caught during the course of a week are given in centimeters. Plot the data on a stem-and-leaf plot: 13, 21, 16, 27, 15, 25, 37, 19, 16, 18, 23, 24, 41, 30, 34

16. Make a stem-and-leaf plot of this data: 5, 26, 37, 15, 24, 31, 8, 11, 17, 21, 22, 21, 9, 28, 29, 30, 34, 38

17. Make a stem-and-leaf plot of the following numbers: 1.5, 2.7, 3.8, 0.4, 0.8, 0.1, 1.8, 1.1, 2.7, 2.6, 3.0, 3.1

Histograms

Histograms are special types of bar graphs. The data plotted on them are from frequency tables. A histogram has intervals that are all the same size, and do not have gaps. Take a look and it will make more sense.

Example 6

Make a histogram of the following data:

12, 0, 21, 31, 33, 14, 7, 7, 20, 18, 15, 8, 4, 39, 27, 4

Do this by organizing the data into a table. Choose intervals for the horizontal axis of the histogram that cover all the data. Notice how the lowest number in the set of data is 0 and the highest is 39. Try out these intervals.

Intervals	# of Values in That Interval
0–9	6
10–19	4
20–29	3
30–39	3

LOOK OUT!

It's always a good idea to count the total number of data points and make sure that you have that same number accounted for in your table. Just one of those helpful hints that prevents errors in the future.

So the intervals in the histogram will be 0–19, 20–29, and 30–39.

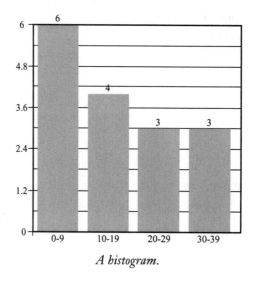

A histogram.

Example 7

A local deli charges the following prices for their lunch specials: $4.25, $7.99, $4.25, $6.85, $10.50, $3.50, $7.99, $5.00, $4.25. Make a histogram of this information.

The most expensive lunch special is $10.50 (you get soup, salad, and a sandwich) and the least expensive one is only $3.50 (you get a bowl of yesterday's soup special and stale bread for that price).

Set your intervals and place the information in a frequency table.

Intervals	# of Values in That Interval
$3–$4	1
$4.01–$5	4
$5.01–$6	0
$6.01–$7	1
$7.01–$8	2
$8.01–$9	0
$9.01–$10	0
$10.01–$11	1

Now plot this information into a histogram.

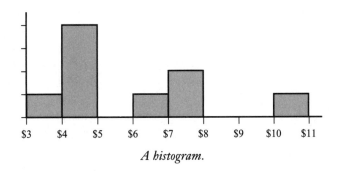

A histogram.

Hopefully you notice a few important things here. First of all, there are several intervals that do not have any values. It is very important that you leave those intervals in the histogram. The histogram should cover *all* intervals.

And there are more lunch specials in the $4.01–$5 range than any other range. Makes budgeting for lunch easier!

Now It's Your Turn

Draw histograms for the following data. You can find the answers in Appendix A.

18. 55, 82, 67, 62, 75, 90, 51, 75, 94

19. 3, 0, 9, 1, 4, 2, 11, 5, 3, 6, 0, 7, 5, 9

20. A survey asked how many phone numbers people had stored in their cell phones. This is the data: 5, 8, 9, 11, 35, 34, 28, 25, 20, 16, 0, 46, 38, 17, 21, 29, 34

21. 0, 8, 5, 9, 17, 16, 15, 14, 10, 12, 21, 22, 20, 18, 18, 18, 19, 27, 28, 22, 20, 23, 29, 26

22. 11, 109, 224, 657, 284, 415, 119, 180, 105, 208

23. 1.24, 2.45, 2.19, 1.99, 1.75, 2.10, 1.65, 2.30, 1.10, 1.11, 2.18

24. 56, 55, 54, 57, 28, 29, 10, 20, 23, 54, 51, 48, 35, 37, 31, 49, 40

25. 1.8, 1.2, 2.1, 2. 8, 3.6, 3.3, 1.8, 2.2, 0.9, 1.0, 1.1, 2.4

Cat in a Container (Box-and-Whisker)

There is one more way to display data to cover in this chapter. This is the box-and-whisker plot. Here you will look at the median, the values above the median, and the values below the median. Most important will be the *upper* and *lower quartiles*.

DEFINITION

An **upper quartile** is the median of the upper half of the data.

A **lower quartile** is the median of the lower half of the data.

Example 8

Make a box-and-whisker plot of these numbers:

172, 209, 176, 260, 200, 197, 254, 236, 295, 186, 266

First thing to do is to order the data from lowest to highest:

172, 176, 186, 197, 200, 209, 236, 254, 260, 266, 295

Then locate the median and the quartiles:

   Lower quartile Median   Upper quartile

  172, 176, 186, 197, 200, **209**, 236, 254, 260, 266, 295

This information will be plotted on a number line, as shown here:

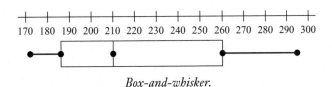

Box-and-whisker.

A vertical line is drawn through the median of the whole set of numbers. A box is drawn from the lower quartile to the upper quartiles. A line segment is drawn from the maximum and another to the minimum. That is a lot of information on one diagram!

Example 9

Make a box-and-whisker plot of these numbers: 60, 15, 12, 25, 19, 22, 55, 72, 41, 42, 30, 37

Order the numbers from lowest to highest: 12, 15, 19, 22, 25, 30, 37, 41, 42, 55, 60, 72

Because there is an even number, the median is the mean of the middle two values. The median is therefore 33.5. The same will be true for the upper quartile and the lower quartile.

Find the mean of the two middle numbers in each case:

  Upper quartile = 48.5

  Lower quartile = 20.5

Plotting all this information on a line graph will look like this:

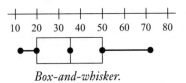

Box-and-whisker.

Now It's Your Turn

Make a box-and-whisker plot of the data provided. The answers are located in Appendix A.

26. 0.8, 0.4, 0.3, 0.2, 0.6, 0.1, 0.7, 0.5, 0.9

29. 60, 62.5, 65, 66, 68, 65, 68, 72, 60, 71.5, 64, 68.5

27. 15, 15, 10, 12, 22, 12, 8, 10, 14, 18, 22, 15, 12, 11

30. 76, 55, 88, 92, 79, 86, 91, 89, 85, 92, 100, 98, 87, 88

28. 38, 58.8, 75, 73, 121, 122, 980

You've Gotta Love Word Problems

In This Chapter

- Solving simple interest and compound interest problems
- Solving area and volume problems
- Using information on rate, distance, and speed to solve word problems
- Setting up and solving mixture or combination problems

Here it is. What you have been waiting for—for at least 15 chapters now. The word problems! Here is how to take what you have learned and really apply it.

Interest(ing) Problems

Interest is money that someone or some institution pays you to use their money. Or it is the money you pay someone or some institution to use their money.

The amount you deposit or borrow is called the principle. The interest rate is the amount that is added to the principle. If interest is paid only on principle, it is called simple interest. Let's solve a simple interest problem.

Example 1

Ryan deposits $250 into a savings account. The annual simple interest rate is 1.5 percent. How much interest will Ryan earn if he leaves the $250 in the bank for 18 months? The formula to find simple interest is $I = Prt$.

The I is the simple interest, P is the principle, r is the annual interest rate, and t is the length of time.

Now, plug in the information that you know. P is $250, the interest rate r is 1.5% = .015, and the length of time is 18 months (or 1.5 years):

$$I = (\$250)(.015)(1.5)$$

$$= \$5.63$$

So after 18 months Ryan has his original $250 plus an additional $5.63 in interest in his account for a total of $255.63.

Example 2

Wilson borrowed $450 from his brother. After nine months he was able to pay back the entire $450 plus $75 in interest. What interest rate did his brother charge him?

Use the same formula:

$$I = Prt$$

This time you are looking for the interest rate, r. Rearrange the equation. Keep in mind that the original principle, P, was $450 and the time is 0.75 (or 9 months out of 12).

$$r = \frac{I}{Pt}$$

$$= \frac{\$75}{450(0.75)}$$

$$= \frac{75}{337.5}$$

$$= .22222 \text{ or } 22\% \text{ interest}$$

Yikes, that is pretty steep. Wilson must have really needed the money.

Compound Interest

Many times someone pays, or is paid, compound interest. This is interest paid on the principle and on the interest earned. This is the type of interest that applies to many savings accounts or retirement funds.

The general formula for compound interest is $A = P(1 + rt)^n$.

Here, P is the principle. The r is the interest rate (as decimal). The t is the length of time in years between compounding. The n is the number of compoundings (or the number of times it has compounded).

PSST—TRY IT THIS WAY

Many examples of compound interest may mature in different ways. For example:

- Monthly means 12 times a year
- Quarterly is 4 times a year
- Semiannually is 2 times a year
- Annually means 1 time a year

Let's work through a problem.

Example 3

Suppose you deposit $1,000 into an account that pays 5 percent annual interest that compounds quarterly. What is the balance at the end of one year? See all the information you have in the problem? That is usually given to you so that you can solve the problem.

This is a rather short example. Let's set up a table to see what is happening here.

Quarter	Beginning Balance	Interest ($I = Prt$)		Ending Balance
1	$1,000	$I = (1,000)(0.05)(0.25)$	$12.50	$1,012.50
2	$1,012.50	$I = (1,012.50)(0.05)(0.25)$	$12.65	$1,025.16
3	$1,025.16	$I = (1,025.16)(0.05)(0.25)$	$12.81	$1,037.97
4	$1,037.97	$I = (1,037.97)(0.05)(0.25)$	$12.97	$1,050.94

The balance after one year is $1,050.94. The total amount of interest that you earned during that year is

$$\$1,050.94 - \$1,000 = \$50.94$$

Money grows slowly.

Example 4

Use the formula for compound interest (instead of the table, although you could do that, too) to find the balance after one year on the following:

You deposit $500 into an account that pays 6 percent interest compounding semiannually.

$$A = P(1 + rt)^n$$

In this case, you have this:

$P = \$500$

$r = 0.06$

$t = 0.5$ (semiannually is twice a year)

$n = 2$ (there are two compounding in a year for a semiannual example)

$A = P(1 + rt)^n$

$= 500[(1 + (0.06)(0.5)]^2$

$= (500)(1.03)^2$

$= 530.45$

The account earned $30.45 in interest that year. That gives you a new balance (at the end of the year) of $530.45.

Now It's Your Turn

Solve the following problems for simple and compound interest. The answers are located in Appendix A.

1. Find the simple interest on $800 at 5 percent for 2 months.

2. Find the simple interest on $25 at 25 percent for 6 months.

3. What is the annual interest rate if you earn $1,600 in interest on an initial investment of $10,000 for one year?

4. Your bank offers an annual interest rate of 6 percent for one year. How much should you deposit into a new account if you want to earn $1,500 in interest during that year?

5. David wants to buy a new flat screen television set. The store is offering a plan which does not charge finance charges if you pay the loan off in one year. Interest compounds monthly. If David buys a television set that costs $1,000 and the store gives him an interest rate of 18 percent, how much interest would accumulate after 6 months?

6. What is the balance on an account when the initial investment is $2,500 and balance compounds monthly at a rate of 9 percent?

7. What is the balance if $2,000 were deposited in an account which has an annual interest rate of 7 percent and is compounded quarterly for 10 years?

Area/Volume Problems

Sometimes mixing in a little geometry makes algebra more applicable to life. Area and volume are expressed by a series of geometric formulas. Don't worry. You don't have to know them off the top of your head. Just make sure you use them correctly and I'll give you them. This is an algebra book after all, not a geometry memorization exercise.

Example 5

Tomato sauce is sometimes sold in cans and sometimes sold in cartons. One rectangular carton is 9 cm long, 3 cm wide, and 17 cm high. A cylindrical can of tomato sauce may be 12 cm high and have a diameter of 9 cm.

Use the following formulas to determine which container holds the most tomato sauce:

volume of a cylinder: $V = bh$

volume of a rectangular prism: $V = lwh$

Start with the rectangular carton. The volume is equal to the length (9 cm) times the width (3 cm) times the height (17 cm):

$V = (9)(3)(17)$

$= 459$ cm^3

The volume of the can is equal to the base times the height.

Here you are looking for the area of the base. The area of the base of a can is the area of a circle. $A = \pi r^2$. So you need to use the radius of the can, and you know the diameter. The radius of a circle $= \frac{1}{2}d$, so:

$r = \frac{1}{2}d$

$= \frac{1}{2}(9)$

The radius of the can is equal to 4.5 cm. This means that the area of the bottom of the can is equal to $A = \pi r^2$. You can use 3.14 as an approximate value for π.

$= (3.14)(4.5)^2$

$= 63.58$ cm^2

Using that information, you can now solve for the volume of the can:

$V = bh$

$= 63.58(12)$

$= 762.96$ cm^3

The can holds nearly one and a half times as much tomato sauce as the carton.

Example 6

A large pizza at Silver Palate Pizza is 16 inches in diameter. Suppose the pizza is cut into 10 equal size pieces. What is the area of each slice of pizza?

First, find the area of an entire 16-inch pizza. The area of a circle is found using the following formula:

$$A = \pi r^2$$

The radius was not included in the information in the problem. But the diameter was. If the diameter of the pizza is 16 inches then the radius is one half of that or 8 inches.

The area of the pizza is

$$A = (3.14)(8)^2$$

$$= 201 \text{ cm}^2 \text{ (approximately)}$$

If the pizza is divided into 10 equal size pieces, then each piece is equal to 20.1 cm².

Now It's Your Turn

Solve each area or volume problem. The answers are located in Appendix A.

8. My new watch has a circular watch face with a radius = 1.75 cm. What is the area of my watch face? (Use 3.14 for the value of π.)

9. A cube-shape puzzle with moveable sides has nine cubes on each side.

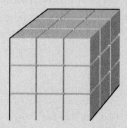

If each cube has a side length of 1.9 cm, what is the surface area of the cube? What is the volume of the whole cube-shape puzzle?

10. Max's birthday present is in a box measuring 40 cm × 25 cm × 4 cm. Should his mother use a piece of wrapping paper that measures 68 cm × 52 cm or 75 cm × 30 cm? Explain.

11. A new supermarket is being built. Its shape and dimensions are given in the diagram. What is the total volume of the building?

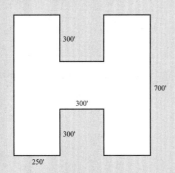

(Figure not shown to scale.)

12. What is the volume of this solid shape?

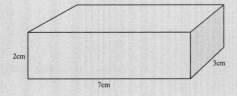

13. Andy has a cylindrical drum from Africa which measures 36 inches high and is 20 inches in diameter. What is the volume of the drum?

14. What is the volume of this cylinder?

15. What is the base area of this cone? What is the volume of the cone?

$$V = \frac{1}{3} \, bh$$

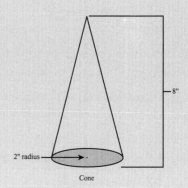

Cone

Distance and Speed Problems

Ever try to figure out how long it is going to take you to get to your destination by car? Ever tried to calculate how long you have run or walked or biked based on how long you have exercised? If so, you were attempting to solve distance and speed word problems. Hopefully this section will make that process a little easier next time.

The generalized formula for distance is $D = rt$ where r is rate and t is time.

Example 7

Jeanne, Alison, Zhanna, and Marta row competitively in a quad scull on the Greenway River. During one practice session they covered the course in approximately 22 minutes, going with the 4 mph current. The trip takes 30 minutes to return against the current. What would their speed be if there were no current? How far is the course, one way?

This is a two-part question. The first thing you want to find is the speed of the boat and rowers if there were no current.

There are two basic equations here. There is the downstream rate and the upstream rate.

The formula for downstream is:

$$D_d = (boat + water) \times T_d$$

The formula for upstream is:

$$D_u = (boat - water) \times T_u$$

Here are the variables in this problem: D_d. Actually, you haven't specifically been told that, but you do know that $D_d = D_u$ because they traveled the same distance upstream as they did downstream. What about the speed of the boat—you don't know that either. What do you know? You know the speed of the water is 4 mph.

$$T_d = 22 \text{ minutes}$$
$$T_u = 30 \text{ minutes}$$

The first thing you want to solve for is their speed when there is no current. To do this, set the two equations equal to each other (because the distance is the same):

$$(boat + water) \times T_d = (boat - water) \times T_u$$

Remember that the times listed are in minutes. You need to convert that to hours. 30 minutes is equal to 0.5 hours and 22 minutes is 0.366 hours.

$$(b + 4)0.366 = (b - 4)0.5$$
$$0.366b + 1.46 = 0.5b - 2$$

Solve for *b*:

$b = 25.8$ mph

Without the current, the boat would travel at 25.8 mph. Now solve for distance. You only need to do it once as the distance upstream and downstream is the same.

$$D_d = (\text{boat} + \text{water}) \times T_d$$

$$= (25.8 + 4) \times 0.366$$

$$= 10.9 \text{ miles}$$

Example 8

Pauline is a long-distance runner. She starts out at an average rate of 8 miles per hour. She runs for a distance and then turns around, following the same route. Her trip home averages 6 miles per hour. If her whole trip takes 2 hours and 15 minutes, how far did she run before she turned around?

What do you know here? You know that the distance out and back on Pauline's run is the same. This means that the distance out = distance home. Distance, remember, is equal to rate times time. The distance out was $8t$ (you know it is at a rate of 8 mph but you don't know the time it took) the distance home is $6(135 - t)$.

What does this mean? Pauline was gone for a total 135 minutes. That is how long she was gone when she arrived home. So it took her $135-t$ (the time it took her to go before she turned around to make the return trip).

So distance out = distance home:

$$8t = 6(135 - t)$$

$$8t = 810 - 6t$$

$$t = 0.96 \text{ hours}$$

Use that to plug into the original equations. Her distance out was $8t$:

$$8(0.96) = 7.7 \text{ miles}$$

Pauline ran 7.7 miles, almost 8 miles, before she came home.

Now It's Your Turn

Solve the following distance and rate problems. The answers are in Appendix A.

16. Sonny travels at an average rate of 67 miles per hour and completes his trip in 3 hours. How much longer would the trip have taken if he traveled at an average rate of 55 miles per hour?

17. Train A and Train B (you had to expect there was going to be at least one problem involving a train leaving a station) leave the station at the same time. Train A is traveling due east and Train B is traveling due west. Train A travels at a rate of 55 mph and Train B travels at a rate of 45 mph. How long will it take for the trains to be 165 miles apart?

18. An airplane makes the 1,400 mile trip from Boston, Massachusetts to Omaha, Nebraska in 3 hours and 25 minutes. For the first part of the trip, the average speed was 110 miles per hour. For the second part of the trip, the average speed was 120 miles per hour. How long did the plane travel at each speed?

19. Two runners start at opposite ends of the bike trail. The trail is 7 miles long. One runner averages 5 miles per hour. The other runner averages 7 miles per hour. If they start at the same time, how long will it take for them to meet?

20. A 747 can travel 1,600 miles in 8 hours if it is traveling with the wind. If the plane is traveling against the wind, it takes 9.25 hours. What is the velocity of the wind?

21. Here is another train one … Train X and Train Y are on parallel tracks. Train X leaves the station at noon, traveling east at 100 mph. Train Y leaves the station at 12:15 P.M. at 115 mph. How long will it take before Train Y catches up with Train X?

22. Eric leaves for work every morning at 7:20 A.M. He drives to the train station (another train!) in his car at an average rate of 35 mph. At the train station, he boards a train which travels at an average rate of 55 mph. At his stop, he leaves the train and walks 10 minutes to his office. Eric walks at a pace of 3 mph. The whole trip covers 48 miles and takes him 1 hour and 15 minutes. What is the distance he travels on the train? Assume that he is able to race from his car to the train without spending an appreciable amount of time parking or waiting.

23. Elizabeth and Louise live 70 miles apart. They agree to meet for lunch at a restaurant which is exactly 35 miles from each of them. Elizabeth travels at a rate of 70 mph and Louise, who has less of a lead foot, travels at a saner 60 mph. How long will it take each to reach their destination?

Mixture and Combination Problems

These problems are just as they sound. They are problems that involve mixing or combining two or more things. It can be a recipe, the water needed to fill a salt water fish tank, or the mixture that is added to a gasoline tank to keep the engine running smooth. It can be a lot of things.

The basic formula that you need to know is given here:

Ingredient a × its value × ingredient b × its value x … = total x its value

Let's take a look at a few examples.

Example 9

Darena has 50 ounces of a 25 percent salt solution. How much water would need to evaporate to make it a 35 percent salt solution? In this case, let's make the original solution Ingredient A in the formula.

You have 50 ounces of 25 percent salt solution. But you aren't looking for salt here; you really want to know about the water. After all, the water is what is evaporated from the scenario.

If there is 25 percent salt in a solution then 75 percent of the solution is water. The 35 percent solution left after evaporation has an unknown amount of water, x, and 65 percent water.

Plug what you know into the equation:

Original solution × its value × unknown amount × its value = final solution × its value

$0.75 \times 50 \times -x = 0.65 \times (50 - x)$

$37.5 - x = 0.65 \times (50 - x)$

$37.5 - x = 32.5 - 0.65x$

Simplify and solve for x:

$5 = 0.35x$

$x = 14.28$

So about 14 ounces of water must evaporate in order for it to be a 35 percent salt solution. Just for fun, let's look at one more example.

Example 10

Forget about evaporating the water, what if you want to just add water? There is a saline solution that is 10 percent salt. How much water needs to be added to make 10 gallons of 5 percent saline solution? You do not know the number of gallons of saline solution to begin with here. Let's give that the variable s. You can now rewrite the problem in terms of s. What is the value of s so that 10 percent of s will be 5 percent of 10 gallons? Hmmmm.

10% of s is $0.10s$

That is your starting ingredient. Your final ingredient is $0.05s$ and 10 gallons. Let's use the formula:

Original solution × its value × unknown amount × its value = final solution × its value

$0.10s \times (10 - s) = 0.05s \times 10$

$s = 5$

So 5 gallons of water would be needed to make 10 gallons of 5 percent saline solution.

Now It's Your Turn

Solve the following word problems. You can find the answers in Appendix A.

24. You want to make 10 gallons of a solution that contains 20 percent alcohol. How many gallons of a 10 percent alcohol solution and a 40 percent alcohol solution must be mixed to achieve this?

25. The Morning Blend coffee sold at the local coffee shop is a mixture of two different coffees. The Morning Blend is made of 5 pounds of coffee that sells for $8.75 per pound and 15 pounds of coffee that costs $5.25 per pound. How much does a pound of the Morning Blend cost?

26. The Patel's are making a drink for their son's birthday party. They make 50 liters of punch. 35 liters is made of a juice that is 15 percent real fruit juice. The rest is made of a juice that is only 5 percent real juice. What is the percentage of real fruit juice in their punch?

27. A sample of a tin and copper alloy weighs 80 pounds. The amount of tin was 12 pounds less than $\frac{2}{3}$ of the copper. How many pounds of tin were there?

28. A 3-pound jar of mixed cocktail nuts has 25 percent almonds. This is mixed with 7 pounds of a different type of mixed cocktail nuts that has 15 percent almonds. What percent of the new mixture are almonds?

29. 10 fluid ounces of a 4 percent alcohol solution is mixed with 25 fluid ounces of a 50 percent alcohol solution. What is the concentration of the new solution?

30. A soil scientist examines a 3m³ sample of soil that has 45 percent clay. She then mixes that soil with a 2m³ sample that is 15 percent clay. What is the clay content of the new mixture?

Answer Key

Chapter 1

1. 40
2. 7
3. 63
4. 14
5. −63
6. $-\dfrac{7}{14}$ or $-\dfrac{1}{2}$
7. 64
8. 3
9. 4
10. 4
11. 3
12. 16.
13. 20
14. 50
15. 56
16. 4
17. −14
18. 15
19. 3
20. −3
21. −1
22. 0
23. Commutative property for multiplication

24. Inverse property for multiplication
25. Identity property for addition
26. Associative property for addition
27. Associative property for multiplication
28. Commutative property for addition
29. Inverse property for addition
30. Identity property for multiplication

Chapter 2

1. $\dfrac{5}{6}$
2. $\dfrac{1}{3}$
3. $\dfrac{1}{7}$
4. $\dfrac{5}{43}$ (sometimes a fraction is in its simplified form)
5. $\dfrac{2}{3}$
6. $\dfrac{3}{8}$
7. $\dfrac{11}{12}$
8. 6
9. $\dfrac{1}{3}$
10. 1
11. $\dfrac{7}{9}$

12. $-1\frac{1}{14}$

13. $\frac{37}{40}$

14. $-\frac{16}{15}$

15. $-\frac{5}{4}$

16. $\frac{11}{9}$

17. $\frac{1}{2}$

18. $\frac{7}{24}$

19. $\frac{17}{6}$

20. $\frac{11}{30}$

21. $\frac{25}{36}$

22. $\frac{2}{3}$

23. $\frac{1}{14}$

24. $\frac{1}{3}$

25. $\frac{3}{16}$

26. $\frac{2}{15}$

27. 1

28. $\frac{5}{18}$

29. $\frac{1}{4}$

30. $\frac{2}{21}$

Chapter 3

1. 100

2. 729

3. 625

4. $\frac{16}{625}$ or 0.0256

5. 1

6. $\frac{1}{65,536}$

7. $-\frac{1}{125}$

8. 15,625

9. 1,024

10. 7^3

11. 1

12. 9.75×10^5

13. 4.78×10^{-6}

14. 5.758×10^7

15. 1.0101×10^{-2}

16. 0.006

17. 4,000,000

18. 900,480,000

19. 0.0005487

20. 42

21. 5

22. 604

23. −33

24. 24

25. 32

26. 29

27. 12

28. 117

29. 0

30. 9

Chapter 4

1. $y = 27$

2. $t = -4$

3. $x = -7$

4. $y = 252$

5. $a = 0.5$

6. $y = 5$

7. 12 feet

8. 4 scarves

9. 39 classes

10. 10 times

11. $\dfrac{23}{9}$

12. 24

13. 11.25

14. $\dfrac{1}{4}$

15. $b = -3$

16. $x = -8$

17. $w = -23$

18. $t = 3.5$

19. Both sides need to be multiplied by 6. The correct answer is $\dfrac{31}{2}$.

20. The sum of the numbers with variable s on the left side is $3s$ not $4s$. The answer is $s = 3$.

21. No solution.

22. $y = 1$ or $y = -15$

23. $p = 0.5$ or $p = -0.5$

24. $\dfrac{3}{2}$

25. 108 cookies (enough for the bake sale and to have some left over!)

26. 24 pages

27. 40 runs

28. 2.5 miles

29. 215

30. $10.44

Chapter 5

1.

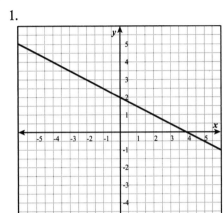

2.

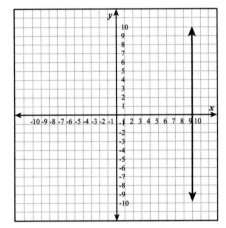

3.

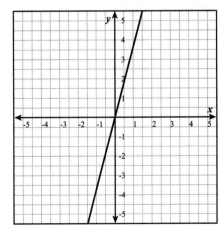

4.

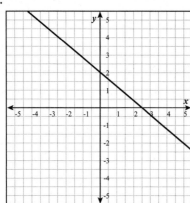

5.

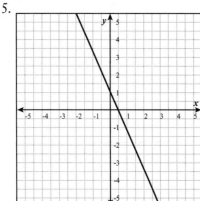

6.

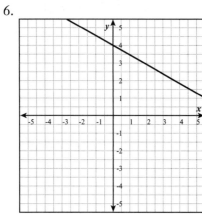

7.

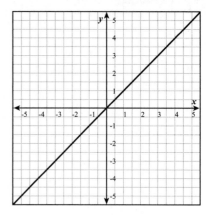

8.

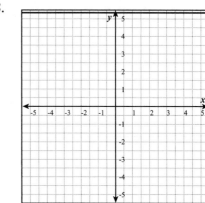

9. $y = x$

10. $y = -x$

11. x-intercept $(\frac{3}{2}, 0)$, y-intercept $(0, -3)$

12. x-intercept $(-\frac{2}{3}, 0)$, y-intercept $(0, -2)$

13. x-intercept $(7, 0)$, y-intercept $(0, 7)$

14. x-intercept $(-\frac{4}{3}, 0)$, y-intercept $(0, 4)$

15. *x*-intercept (6,0), *y*-intercept (0,7)

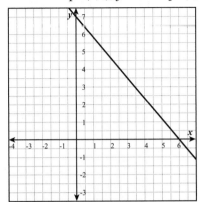

18. *x*-intercept (50,0), *y*-intercept (0,–10)

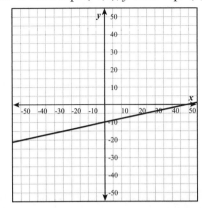

16. *x*-intercept (2,0), *y*-intercept (0,–2)

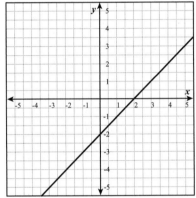

19. *x*-intercept (6,0), *y*-intercept (0,–6)

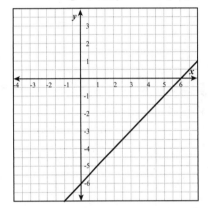

17. *x*-intercept $(-\frac{1}{3}, 0)$, *y*-intercept $(0, \frac{5}{3})$

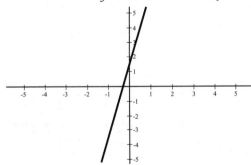

20. *x*-intercept (4,0), *y*-intercept (0,–12)

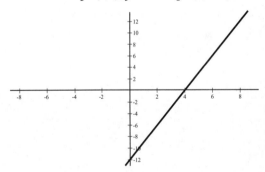

21. $m = -7$

22. $m = \frac{1}{5}$

23. $m = -3$

24. $m = \dfrac{3}{2}$

25. The answer subtracts the values for x in the numerator and the values of y in the denominator rather than the other way around. The answer should be 1.

26. The answer sets up the numerator so that it is $y_2 - y_1$ but the denominator has $x_1 - x_2$. The correct answer is 1.

27. $\dfrac{1}{2}$

28. –14 customers per week

29. 36 customers per week

30. Between weeks 1 and 4

Chapter 6

1. $y - 4 = 2(x + 1)$

2. $y - 11 = -6(x + 3)$

3. $y - 5 = \dfrac{3}{4}(x - 7)$

4. $y - 4 = -(x + 5)$

5. $y + 2 = \dfrac{2}{3}(x - 3)$

6.

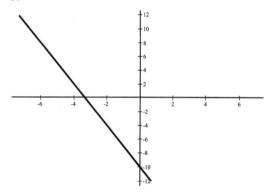

Graph of $y + 4 = -3(x + 2)$.

7.

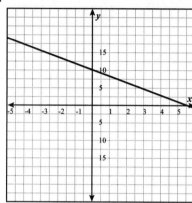

Graph of $y - 5 = -2(x - 3)$.

8.

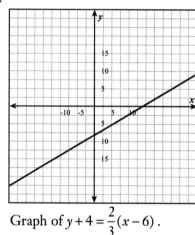

Graph of $y + 4 = \dfrac{2}{3}(x - 6)$.

9.

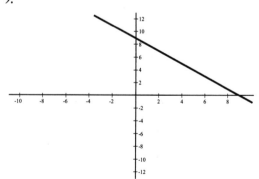

Graph of $y - 3 = -(x - 6)$.

10.

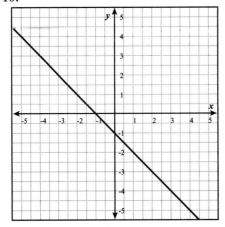

Graph of $y - 1 = -(x + 2)$.

11. $m = -2, b = 0$

12. $m = 3, b = -5$

13. $m = -\dfrac{1}{8}, b = -3$

14. $m = 2, b = \dfrac{2}{3}$

15. $m = 1, b = 9$

16. $m = -6, b = 12$

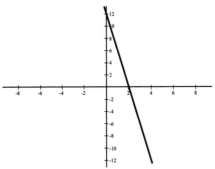

17. $m = -\dfrac{1}{4}, b = 8$

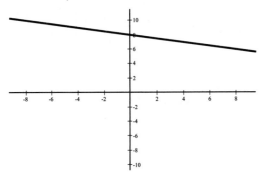

18. $m = -\dfrac{2}{3}, b = 1$

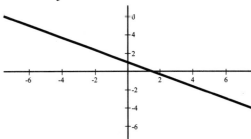

19. $m = -1, b = 0$

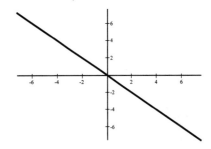

20. $m = 3, b = -6$

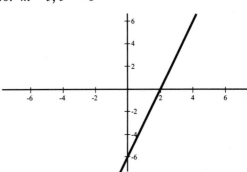

21. Perpendicular

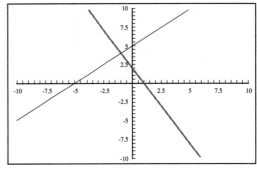

Legend
$y = x + 5$
$4x + 2y = 4$

Graph of $y = x + 5$ and $4x + 2y = 4$.

22. Neither, these lines are the same.

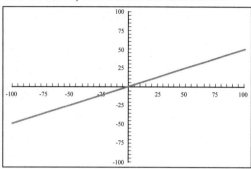

Graph of $-x + 2y = 1$ and $y = \dfrac{1}{2}x + \dfrac{1}{2}x$.

23. Parallel

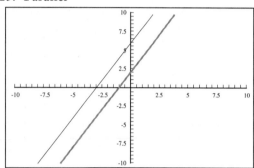

Legend

$y = 2x + 6$

$y = 2x + 2$

Graph of $y = 2x + 6$ and $y = 2x + 2$.

24. Neither, the lines are the same.

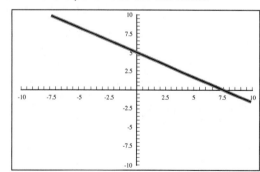

Legend

$2x + 3y = 15$

$y = -\dfrac{2}{3}x + 5$

Graph of $2x + 3y = 15$ and $y = -\dfrac{2}{3}x + 5$.

25. Parallel

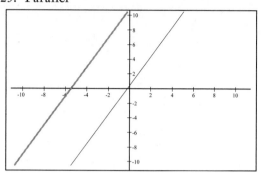

Legend

$2y - 4x = 1$

$y = 2x + 11$

Graph of $2y - 4x = 1$ and $y = 2x + 11$.

26.

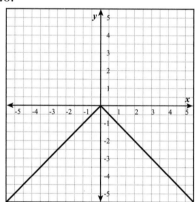

Graph of $f(x) = -|x|$.

27.

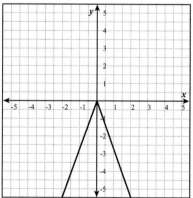

Graph of $g(x) = -3|x|$.

28.

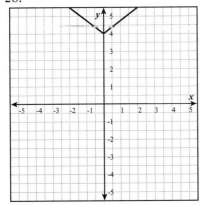

Graph of $f(x) = |x| + 4$.

29.

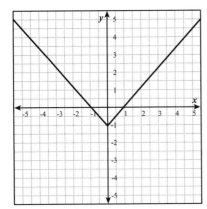

Graph of $g(x) = |x| - 1$.

Chapter 7

1. $y = 5x - 3$
2. $y = -x + 6$
3. $y = \frac{1}{5}x + 8$
4. $y = -4x - \frac{2}{7}$
5. $y = 3x + 3$
6. $y = -6x + 1$
7. $y = \frac{2}{3}x$
8. $y = -\frac{5}{2}x + 4$
9. $y = 3x + 5$
10. $y = \frac{1}{4}x + 1$

11. $y = -x + 6$
12. $y = \frac{1}{3}x - 5$
13. $y = -\frac{3}{5}x - 6\frac{1}{5}$
14. $y = -3x - 2$
15. $y = x + 2$
16. $y = -6$
17. $2x + y = 8$
18. $\frac{2}{3}x + y = -\frac{2}{3}$
19. $\frac{3}{5}x + y = -\frac{4}{5}$
20. $-\frac{3}{4}x + y = \frac{3}{4}$
21. Parallel, they have the same slope
22. No, their slope is not the negative reciprocal
23. $y = \frac{2}{5}x + 5$
24. $y = -\frac{5}{2}x + \frac{39}{2}$
25. No, they do not have the correct slopes
26. $y = .226x + 10.14$
27. $y = 2.03x + 7.73$
28. $y = .464x + 4$
29. $y = -3.54x + 13.71$
30. $y = 31.29x - 12.43$

Chapter 8

1. $y \le 1$
2. $x > \frac{9}{4}$
3. $y \ge -\frac{5}{3}$
4. $p \ge -\frac{10}{3}$
5. $y \le \frac{7}{18}$

6. The solution is the set of all real numbers.

7. $b < 10$

8. There is no solution.

9.
Graph of $3y + 9 \leq 12$.

10.
Graph of $4x - 2 > 7$.

11.
Graph of $18 - 6y \geq 28$.

12.
Graph of $4p - 3 \leq 7p + 7$.

13.
Graph of $-3(6y + 5) \geq -22$.

14.
Graph of $4y - 5 > 3y + y - 7$.

15.
Graph of $3(b + 6) < 48$.

16.
Graph of $5(t-6) > 4t - 16$.

17.
Graph of $3 < x + 7 < 9$.

18.
Graph of $-8 \leq y - 5 < 12$.

19.
Graph of $4x + 6 < 18$ or $6x - 12 > 24$.

20.
Graph of $4n + 2 < -9$ or $3n - 7 > 7$.

21.
Graph of $|y + 5| < 8$.

22.
Graph of $4|2z - 3| - 5 < 11$.

23.
Graph of $|r + 8| \geq 19$.

24.
Graph of $|5b - 7| \leq 20$.

25.
Graph of $|b| \geq -9$.

26.
Graph of $4 + m \geq 0$.

27.
Graph of $y < x - 2$.

28.
Graph of $y \geq -3x + 7$.

29.

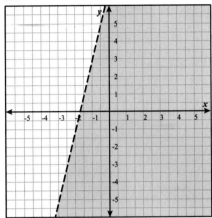

Graph of $y < 4x + 7$.

30.

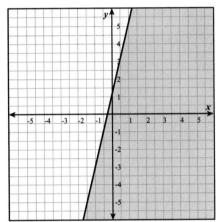

Graph of $4x - y \geq -1$.

Chapter 9

1. Solution $(-7.6, -4.3)$

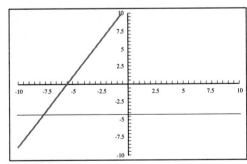

Graph of $2x - y = -11$ and $y = -2y - 13$.

2. Solution $(5,4)$

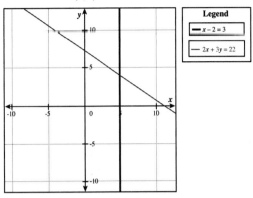

Graph of $x - 2 = 3$ and $2x + 3y = 22$.

3. Solution $(0,5)$

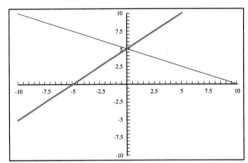

Graph of $x + 2y = 10$ and $-x + y = 5$.

4. $(1.11, -5.444)$

5. $(\frac{21}{11}, \frac{18}{11})$

6. $(\frac{21}{2}, -12)$

7. $(10, -15)$

8. $(-\frac{45}{14}, \frac{9}{14})$

9. $(\frac{100}{29}, \frac{252}{29})$

10. $(\frac{213}{64}, \frac{35}{16})$

11. One solution, $(1, -\frac{9}{4})$

12. One solution $(17, 11)$

13. No solution

14. Infinite number of solutions

15. One solution $(\frac{5}{3}, 0)$

16. One solution (0,13)

17. No solution

18. No solution

19. One solution $(\frac{36}{11}, \frac{48}{11})$

20. One solution (–1,0)

21.

Graph of $x < 5$ and $x > -2$.

22.

Graph of $y \leq 12$ and $y \geq 8$.

23.

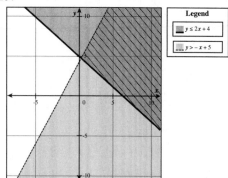

Graph of $y \leq 2x + 4$ and $y > -x + 5$.

24.

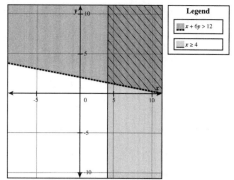

Graph of $x + 6y > 12$ and $x \geq 4$.

25.

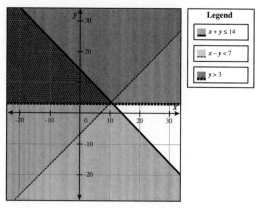

Graph of $x + y \leq 14$ and $x - y < 7$ and $y > 3$.

26.

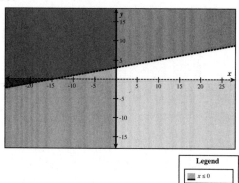

Graph of $x \leq 0$ and $y < 0$ and $5y - x > 15$. The area shaded with diagonal lines is the solution.

27.

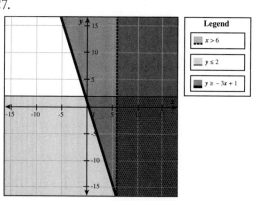

Graph of $x > 6$ and $y \leq 2$ and $y \geq -3x + 1$.

28.

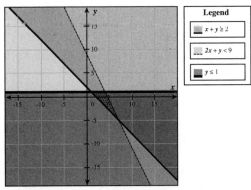

Graph of $x + y \geq 2$ and $2x + y < 9$ and $y \leq 1$.

29.

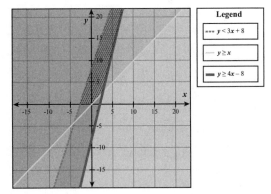

Graph of $y < 3x + 8$ and $y \geq x$ and $y \geq 4x - 8$.

30.

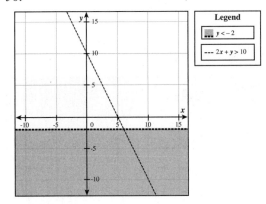

Graph of $y < -2$ and $2x + y > 10$.

Chapter 10

1. $\begin{bmatrix} 2 & 8 \\ 3 & -6 \end{bmatrix}$

2. $\begin{bmatrix} -8 & 14 \\ -2 & 4 \end{bmatrix}$

3. You cannot add these two because they do not have the same dimension.

4. $\begin{bmatrix} 8 & 4 & -1 \\ -5 & -12 & -11 \\ -7 & 18 & 5 \end{bmatrix}$

5. $\begin{bmatrix} -2 & -5 & 13 \\ 1 & 19 & -10 \\ 10 & 3 & 9 \end{bmatrix}$

6. $\begin{bmatrix} 6 & 24 & 7 & -23 \\ 10 & 21 & -5 & 12 \end{bmatrix}$

7. $\begin{bmatrix} 11 & 13 \\ -9 & -9 \\ 14 & 2 \end{bmatrix}$

8. $\begin{bmatrix} 5 & 13 & 9 & 5 \\ 19 & -4 & 12 & -26 \\ 9 & -3 & -11 & 6 \\ 8 & 8 & -5 & -6 \end{bmatrix}$

9. $\begin{bmatrix} 0 & 8 \\ 0 & 12 \end{bmatrix}$

10. $\begin{bmatrix} 45 & -5 \\ 20 & -35 \end{bmatrix}$

11. $\begin{bmatrix} -4 & -36 & 62 \\ -30 & 38 & 47 \\ 43 & -30 & 65 \end{bmatrix}$

12. $\begin{bmatrix} -81 & -51 \\ 38 & 7 \end{bmatrix}$

13. This problem has invalid dimensions and can't be solved.

14. $\begin{bmatrix} 63 & -54 \\ -7 & 10 \\ 28 & -44 \end{bmatrix}$

15. $\begin{bmatrix} 9 & 252 \\ -5 & -22 \\ 24 & 82 \end{bmatrix}$

16. $\begin{bmatrix} 42 & 72 \\ -36 & 0 \\ -24 & -30 \end{bmatrix}$

17. $\begin{bmatrix} 738 & -858 \\ 108 & 180 \\ -288 & 390 \end{bmatrix}$

18. $\begin{bmatrix} -18 & -30 \\ 72 & -54 \end{bmatrix}$

19. 31

20. −29

21. −22

22. 382

23. 154

24. 1661

25. $x = -\dfrac{46}{9}, \quad y = \dfrac{22}{9}$

26. $x = -\dfrac{4}{3}, \quad y = \dfrac{14}{3}$

27. $x = -\dfrac{89}{26}, \quad y = -\dfrac{75}{26}$

28. $x = -1, \ y = 3$

Chapter 11

1. $10b^2 + 3$

2. $-3x^2 - 5x + 1$

3. $-4y^3 - 7y^2 + 6y - 22$

4. $12r^3 + 3r^2 + 19r - 33$

5. $25x^2 - 20x + 4$

6. $8x^3 + 20x^2 - 24x + 6$

7. $3x^3 + 15x^2 + 13x + 2$

8. $8w^2 + 5w - 3$

9. $20a^2 + 16a - 36$

10. $(t - 5)(t + 4)$

11. $(x - 7)(x - 3)$

12. $(w - 6)(w + 4)$

13. $(3m - 5)(m + 1)$

14. $(4x - 1)(x + 7)$

15. $(-3g + 9)(g - 1)$

16. $(2y - 7)(2y + 7)$

17. $(3w - 9)(3w + 9)$

18. $(4r + s)^2$

19. $(t + 3)^2$

20. $2(p - 5)^2$

21. $(x^2 + 1)(x + 4)$

22. $(a + 5)(a + b)$

23. $(7y^2 - 6)(y - 4)$

24. $(r - 3)^2$

25. $3y(y + 4)(y - 1)$

26. $(a - 10)(b - 2)$

27. $4y(y^2 + 4y + 8)$

28. $x(x + 7)(x - 7)$

29. $4(2x^3 - x^2y - 8y + 4)$

30. $(x + 8)(x - 3)$

Chapter 12

1. $\dfrac{\sqrt{5}}{5}$

2. $\dfrac{1}{3}$

3. $2\sqrt{6}$

4. $6\sqrt{3}$

5. $\dfrac{4\sqrt{7}}{7}$

6. $9\sqrt{3}$

7. $-5\sqrt{3}$

8. $-4\sqrt{6}$

9. $4\sqrt{4} + 24\sqrt{3}$

10. $\sqrt{5} + 45\sqrt{2}$

11. $7\sqrt{6} - 6\sqrt{2}$

12. $6\sqrt{3} + 2\sqrt{2}$

13. $3\sqrt{2} + 3\sqrt{3}$

14. 15

15. $12\sqrt{2} - 8$

16. $4\sqrt{5} + 4\sqrt{7} + 2\sqrt{10} + 2\sqrt{14}$

17. $\dfrac{\sqrt{5x}}{15}$

18. $\dfrac{2}{y}$

19. $\dfrac{2\sqrt{15}}{15}$

20. $x = 4$

21. $x = 28.8$

22. $x = 18$

23. $x = 1$

24. No real solution.

25. $y = -\dfrac{19}{9}$

26. $x = -\dfrac{18}{7}$

27. $x = -1.69, x = 2.68$

28. $x = 7.03$

29. No real solution.

Chapter 13

1. Domain: 3, 2, 5, –2

    Range: 10, –1, 8, –1, 2

    Not a function

2. Domain: 0, 1, 3, 8, 9

    Range: 5, 6, 7, 1, 0

    Function

3. Domain: –4, –1, 3, 6

    Range: 2, 5, 6, 2

    Function

4. Domain: –1, 3, 4, –2

    Range: –1, 5, 8, 0

    Function

5.

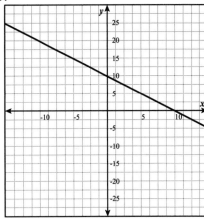

Graph of $y = -x + 10$.

6.

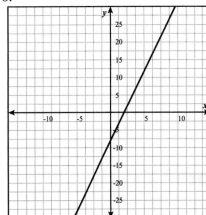

Graph of $y = 4x - 8$.

7.

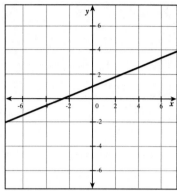

Graph of $y = \dfrac{2}{5}x + 1$.

8.

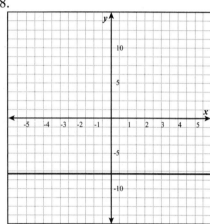

Graph of $y = -8$.

9. Linear

10. Not linear

11. $y = \dfrac{1}{4}x + \dfrac{5}{4}$

12. $y = -\dfrac{1}{5}x - \dfrac{1}{5}$

13. $y = \dfrac{3}{x}$

14. $y = 3x - 5$

15. $y = x - 6$

16. $y = 3x + 15$

17. $y = \dfrac{13}{x} + 2$

18. $y = \dfrac{x - 1}{6}$

19. $y = \dfrac{3x - 12}{2}$

20. $y = -x$

21.

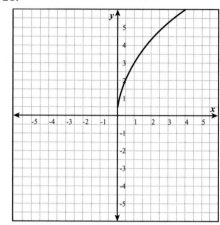

Graph of $y = 3\sqrt{x}$.

22.

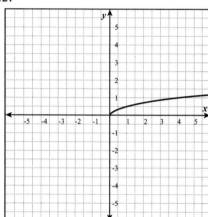

Graph of $y = \dfrac{1}{2}\sqrt{x}$.

23.

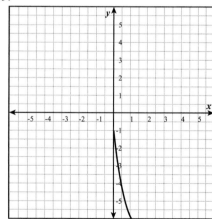

Graph of $y = -6\sqrt{x}$.

24.

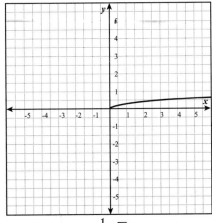

Graph of $y = \dfrac{1}{4}\sqrt{x}$.

25.

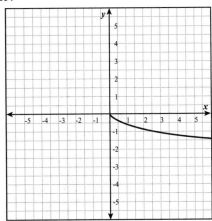

Graph of $y = -0.6\sqrt{x}$.

26.

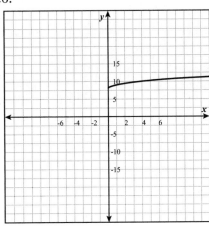

Graph of $y = \sqrt{x} + 8$.

27.

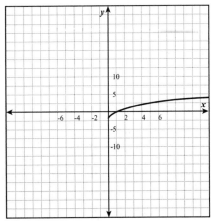

Graph of $y = \sqrt{x} - 1$.

28.

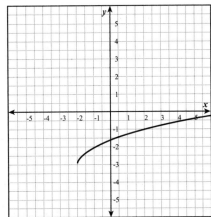

Graph of $y = \sqrt{x+2} - 3$.

29.

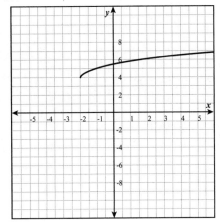

Graph of $y = \sqrt{x+2} + 4$.

30.

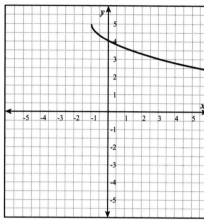

Graph of $y = -\sqrt{x+1} - 5$.

Chapter 14

1.

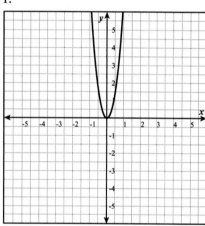

Graph of $y = 7x^2$.

2.

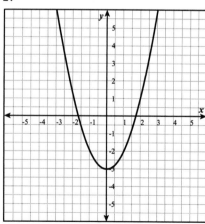

Graph of $y = x^2 - 3$.

3.

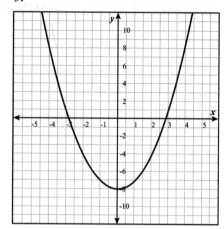

Graph of $y = x^2 - 8$.

4.

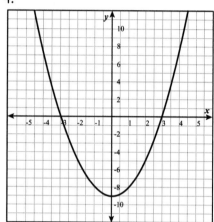

Graph of $y = x^2 - 9$.

5.

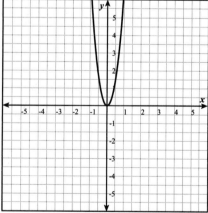

Graph of $y = \dfrac{13}{2}x^2$.

6. Axis of symmetry $x = \dfrac{3}{4}$; vertex $= (\dfrac{3}{4}, 7\dfrac{7}{8})$

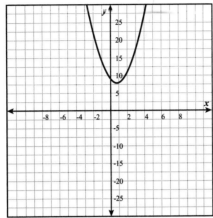

Graph of $y = 2x^2 - 3x + 9$.

7. Axis of symmetry $x = 2$; vertex $= (2, 16)$

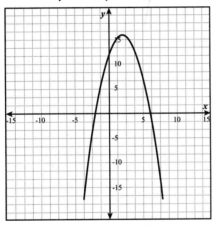

Graph of $y = -x^2 + 4x + 12$.

8. Axis of symmetry $x = 2$; vertex $= (2, 16)$

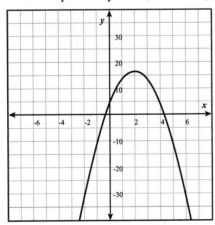

Graph of $y = -3x^2 + 12x + 4$.

9. Axis of symmetry $x = -\dfrac{1}{2}$; vertex $= (-\dfrac{1}{2}, -\dfrac{9}{4})$

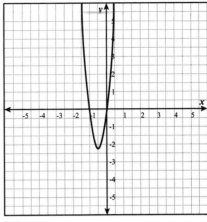

Graph of $y = 9x^2 + 9x$.

10. Axis of symmetry x $= -10$; vertex $=$ $(-4.5, -11.76)$

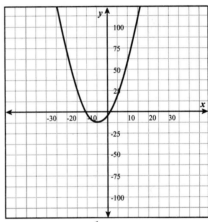

Graph of $y = \dfrac{1}{3}x^2 + 3x - 5$.

11.

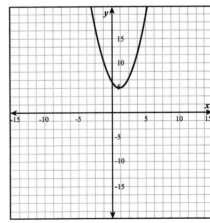

Graph of $y = x^2 - 2x + 6$. No solution.

12.

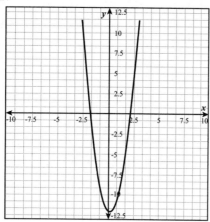

Graph of $y = 3x^2 - x - 12$. Solutions $x = 2.12$ and $x = -1.84$.

13.

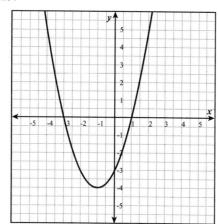

Graph of $y = x^2 + 2x - 3$. Solutions $x = -3$ and $x = 1$.

14. $y = 2 \pm \sqrt{19}$

15. $y = 0$

16. $y \pm \dfrac{1}{2}$

17. $x = -5 + \sqrt{29}$ and $x = -5 - \sqrt{29}$

18. $x = 0$ and $x = 4$

19. $r = 7 + \sqrt{2}$ and $x = 7 \pm \sqrt{47}$

20. $x = 2, x = -12$

21. $r = 6 + \sqrt{30}, r = 6 - \sqrt{30}$

22. $x = -4\sqrt{19}$

23. $x = -4.3, x = 2.4$

24. $x = -4, x = -2$

25. Two solutions, two x-intercepts

26. Two solutions, two x-intercepts

27. Two solutions, two x-intercepts

28. Two solutions, two x-intercepts

29. Two solutions, two x-intercepts

30. Two solutions, two x-intercepts

Chapter 15

1.

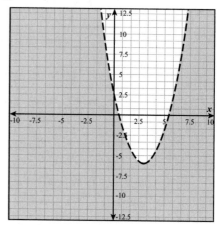

Graph of $y < x^2 - 6x + 3$.

2.

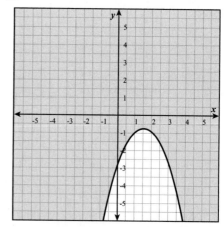

Graph of $y \geq -x^2 + 3x - 3$.

3.

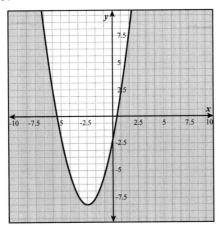

Graph of $y \leq x^2 + 5x - 2$.

6.

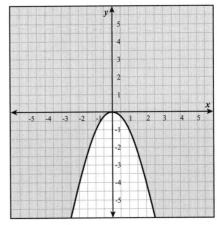

Graph of $y \geq -x^2$.

4.

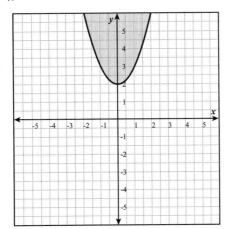

Graph of $y \geq x^2 + 2$.

7.

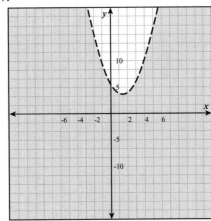

Graph of $y < x^2 - 3x + 6$.

5.

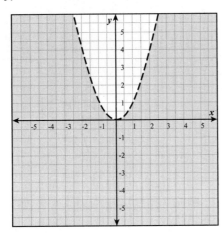

Graph of $y < x^2$.

8.

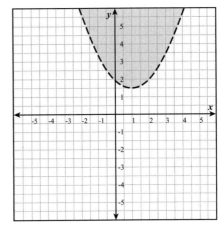

Graph of $y > 0.5x^2 - x + 2$.

9.

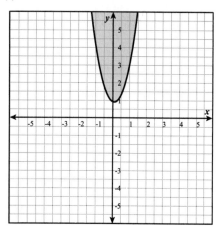

Graph of $y + x \geq 3x^2 + 1$.

10.

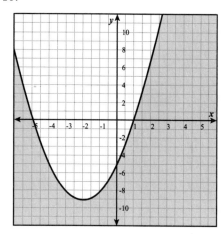

Graph of $y \leq x^2 + 4x - 5$.

11.

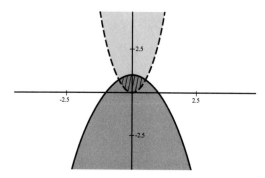

Graph of $y > 3x^2$ and $y \leq -x^2 + 1$.

12.

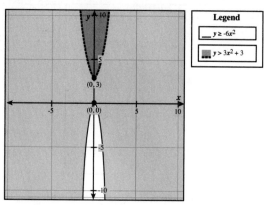

Graph of $y \geq -6x^2$ and $y > 3x^2 + 3$.

13.

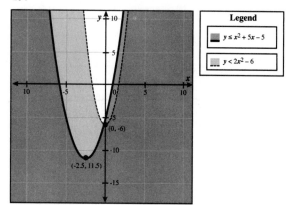

Graph of $y \leq x^2 + 5x - 5$ and $y < 2x^2 - 6$.

14.

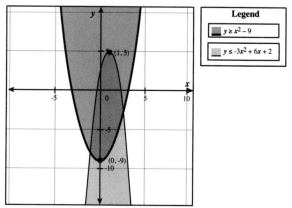

Graph of $y \geq x^2 - 9$ and $y \leq -3x^2 + 6x + 2$.

15.

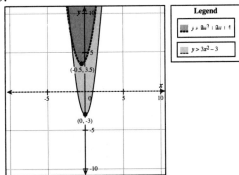

Graph of $y > 2x^2 + 2x + 4$ and $y > 3x^2 - 3$.

16. $-1.42 < x < 8.42$

17. $x < 2.22$ or $x > 4.72$

18. $-3 < x < 1$

19. $x \leq -2.275$ or $x \geq 5.275$

20. $-.89 < x < 1.27$

21. $x \leq -1$ or $x \geq 1.33$

22. $x \leq -2.5$ or $x \geq -.67$

23. $x < .169$ or $x > 11.83$

24. $-2.24 < x < 2.24$

25. $x \leq -5.45$ or $x \geq -.55$

26. $.14 < x < 10.48$

27. $-4 \leq x \leq 4$

28. $x \leq 2$ or $x \geq 6$

29. $-1 \leq x \leq .75$

30. $x < -.56$ or $x > 3.56$

Chapter 16

1. $\dfrac{1}{7x^2}$, excluded value $x = 0$

2. $\dfrac{3y}{y-5}$, excluded value $y = 5$

3. $\dfrac{x}{3}$, excluded value $x = -4$

4. $\dfrac{1}{x^3 + 2x}$, excluded value $x = 0, -2$

5. $\dfrac{m-4}{m-12}$, excluded value $x = 12, -4$

6. $\dfrac{y+1}{y+5}$, excluded value $y = -5, -3$

7. $\dfrac{y+2}{y^2-81}$, excluded value $y = -9, 9$

8. $\dfrac{28x^2 + 24}{12x^2}$

9. $\dfrac{36x}{(x+4)(x-4)}$

10. $\dfrac{2x^2 - 4x + 14}{(x+4)(x+2)(x-2)}$

11. $\dfrac{40-5x}{15x^3}$

12. $\dfrac{-4x^2 + 9x + 63}{(4x)(x+7)}$

13. $\dfrac{7x^3 + 31x + 6}{(x^2+4)(x+2)}$

14. $\dfrac{19y^2 - y - 20}{9y(y-5)}$

15. $\dfrac{-y^2 + 7y - 32}{(y-3)^2(y-8)}$

16. $\dfrac{6x}{x-2}$

17. $\dfrac{6x}{x-2}$

18. $\dfrac{b-2}{5b}$

19. $\dfrac{-3w^2(w+1)}{(w+2)(w+11)}$

20. $\dfrac{x-3}{15x}$

21. $\dfrac{x^2 + 5x}{-x^3 + 2x^2 - 5x + 10}$. The problem is the same as the original. Nothing cancels.

22. $\dfrac{125y^3}{64}$

23. $\dfrac{a^2 - 3a}{-a^2 - 5a + 14}$

24. $\dfrac{1}{2h}$

25. $\dfrac{1}{x-1}$

26. $4(y + 4)$

27. $-117t^6$

28. $\dfrac{f}{2(f-3)(f-9)}$

29. $-\dfrac{1}{18x^3}$

30. $\dfrac{(4x^2 + 7x - 8)(x+1)(x-1)}{(2x-1)(x+3)(5x-1)}$

Chapter 17

1. $x = \dfrac{4}{3}$

2. $x = 0$

3. $y = \dfrac{2}{5}$

4. No real solution

5. $k = 12$

6. $f = 8, f = -6$

7. $m = -2.648, m = -14.35$

8. $x = 10.12, x = -7.117$

9. $x = -6, x = 1$

10. $y = -\dfrac{32}{5}$

11. Direct variation

12. Direct variation

13. Inverse variation

14. Neither

15. Direct variation

16.

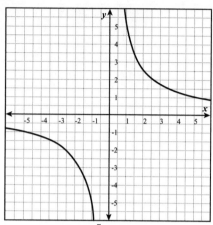

Graph of $y = \dfrac{5}{x}$.

17.

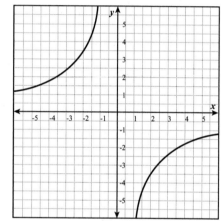

Graph of $y = -\dfrac{7}{x}$.

18.

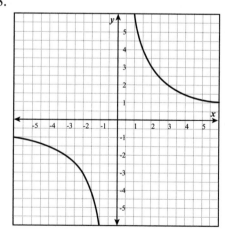

Graph of $2xy = 12$.

19.

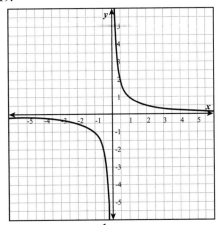

Graph of $y = \dfrac{1}{x}$.

20.

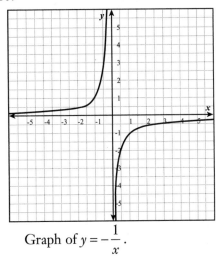

Graph of $y = -\dfrac{1}{x}$.

21.

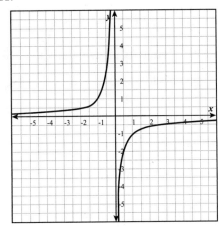

Graph of $xy = -1$.

22.

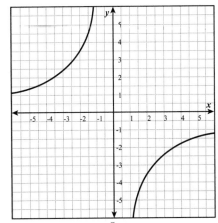

Graph of $y = -\dfrac{5}{x}$.

23.

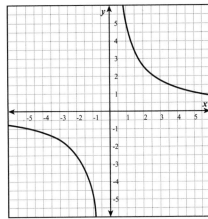

Graph of $2x = \dfrac{10}{y}$.

24. $y = \dfrac{16}{x}$

25. $y = \dfrac{24}{x}$

26. $y = \dfrac{72}{x}$

27. Not an inverse variation

28. $y = \dfrac{-24}{x}$

29. $y = \dfrac{300}{x}$

Chapter 18

1. 4.5%

2. 50%

3. 56%

4. 12.5%

5. $\dfrac{1}{26}$

6. odds $\dfrac{1}{349}$, probability $\dfrac{1}{350}$

7. $\dfrac{9}{7}$ or 9:7

8. $\dfrac{2}{3}$ or 2:3

9. $\dfrac{4}{25}$

10. 3:2

11. 120 different ways

12. 362,880 different ways

13. 40,320

14. 1

15. 1

16. $\dfrac{1}{5,040}$

17. 210 combinations

18. 20,160

19. $\dfrac{1}{120}$

20. 11:16

21. $\dfrac{5}{6}$

22. $\dfrac{2}{3}$

23. $\dfrac{6}{6}$ or 1

24. Dependent $\dfrac{56}{110}$

25. Independent $\dfrac{1}{36}$

26. Independent $\dfrac{1}{8}$

27. Dependent $\dfrac{1}{21}$

28. Dependent $\dfrac{1}{40}$

29. Independent $\dfrac{49}{256}$

Chapter 19

1. Mean 5.1; median 3; mode 0, 1

2. Mean 34; median 34; there is no mode

3. Mean 0.43; median 0.25; there is no mode

4. Mean 79.7; median 78; there is no mode

5. Mean 8; median 7; mode 1

6. She has to get at least a 113. She better hope for lots of extra credit!

7. Range = 7.31; mean absolute deviation = 2.2 (rounded)

8. Range = 34; mean absolute deviation = 10.17

9. Range = 24; mean absolute deviation = 5.9

10. Range = 13; mean absolute deviation = 3.03

11. Range = 9; mean absolute deviation = 2.44

12. Range = 6; mean absolute deviation = 1.5

13.
Stem	Leaves
9	0 3 4 8
10	0 1 2 3
11	1 9
12	1 8

Key: 9|3 = 93

14. Stem	Leaves
3	6
4	5
5	3 3
6	2 6
7	2

Key: 5|3 = 53 years

15. Stem	Leaves
1	3 5 6 6 8 9
2	1 7 5 3 4
3	0 7 4
4	1

Key: 2|1 = 21 inches

16. Stem	Leaves
0	5 8 9
1	1 5 7
2	6 4 1 2 1
3	7 1 0 4 8

Key: 2|6 = 26

17. Stem	Leaves
0.	4 8 1
1.	5 8 1
2.	7 7 6
3.	8 0 1

Key: 3.|0 = 3.0

18.
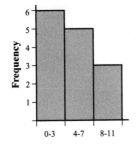

19.
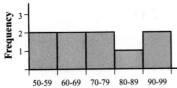

20.
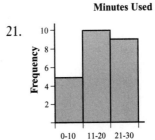

21.
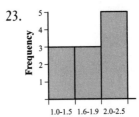

22.
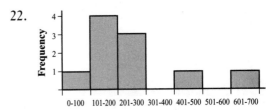

23.
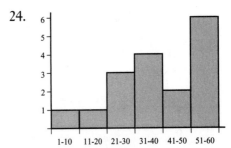

24.
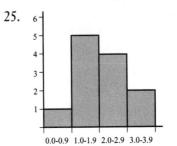

25.
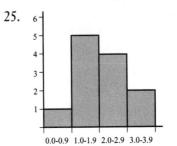

26.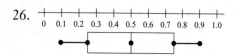

27.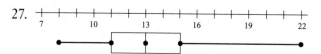

28.

29.

30.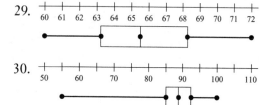

Chapter 20

1. $6.66 in interest

2. $3.12 in interest

3. 16%

4. $25,000

5. $90

6. $2,734.52

7. $4,003.19

8. 9.6 cm²

9. Surface area 194.94 cm²; volume 185.19 cm³

10. The 68 cm × 52 cm sheet of wrapping paper is big enough to cover the box. The other piece does not have enough paper.

11. 114, 380, 000 ft³

12. 42 cm³

13. 11,304 in³

14. 3,846.5 cm³

15. Base area 12.56 in²; volume 33.5 in³

16. 39 minutes longer

17. It would take exactly 1 hour 39 minutes.

18. Approximately 1 hour 49 minutes at 110 mph and 1 hour 26 minutes at 120 mph

19. Exactly 35 minutes

20. 13.5 mph

21. 115 minutes

22. 26.4 miles

23. Elizabeth – 30 minutes; Louise – 35 minutes

24. 3.33 liters of the 40% solution and 6.67 liters of the 10% solution.

25. $6.12

26. 12%

27. 24.8 lb. of tin

28. 18%

29. 36.9%

30. 33%

absolute deviation   The absolute value of the difference between x and a given value, the absolute deviation = $|x - \text{given value}|$.

absolute value   The distance between a number, a, and 0 on a number line. This is represented by $|a|$.

addition property of equality   States that if the same numbers are added to each side of an equation, then the results are equal. If $a = b$, then $a + c$ is also equal to $b + c$.

addition property of inequality   States that adding the same number to each side of an inequality does not change the relationship. If $a > b$, then $a + c$ is still $> b + c$.

associative property of addition   States that changing the sequence of numbers being added does not change the sum. For example, $(a + b) + c = a + (b + c)$.

associative property of multiplication   States that changing the sequence of numbers being multiplied does not change the product. In other words, $(a \times b) \times c = a \times (b \times c)$.

axis of symmetry   The line that passes through the vertex in a graph of a quadratic function. It divides the parabola in half.

binomial   A polynomial with two terms. For example, $x^3 + 6x$ is a binomial.

box-and-whisker plot   A way to display data visually that organizes the data into four groups: minimum value, lower quartile, median, upper quartile, and maximum value.

central tendency   Ways of looking at data including mean, median, and mode of a data set.

coefficient   The part of an algebraic term that is the number; it has a variable associated with it.

commutative property of addition   States that changing the order of numbers being added does not change the sum: $a + b = b + a$.

commutative property of multiplication   States that changing the order of numbers being multiplied does not change the product: $a \times b = b \times a$.

complex fraction   A fraction that has another fraction as part of the numerator, the denominator, or both the numerator and the denominator: $\dfrac{\frac{2}{c^3}}{5}$.

compound event   An event that has two or more events. These are designated with the words *and* or *or*. If you roll a die and flip a coin at the same time, this is a compound event.

compound inequality   Two inequalities joined by the terms *and* or *or*. For example, $-8 > x$ and $x \leq 6$ is a compound inequality.

compound interest   Interest earned on an initial investment and on the interest already earned on the account.

counting principle   States that if there are m ways of making one choice and n ways of making a second choice, then there are $m \times n$ ways of making the first choice followed by the second choice.

Cramer's Rule   A way to solve systems of equations that uses the determinants of a matrix.

cross products property   Used with proportions, the product of the numerator of one ratio and the denominator of the other ratio.

degree of the monomial   The sum of the exponents of the variables of a monomial. The degree of a constant that is not zero is 0.

denominator   The bottom part of a fraction that acts as the divisor.

dependent compound event   Two events in which the occurrence of one event impacts the other event.

dimensions   The number of rows and columns in a matrix, written as $m \times n$.

discriminant   The expression under the radical sign in the quadratic formula. That is the expression $b^2 - 4ac$.

distance   The distance between any two points (x_1, y_1) and (x_2, y_2) $= \sqrt{(x_2 - x_1)^2 + (y_2 - y_1)^2}$; also the product of rate and time in a word problem.

distributive property   A property that can be used to find the product of a number and a sum or difference. The property says: $a(b + c) = ab + ac$ and $a(b - c) = ab - ac$.

domain   The set of all inputs in a function.

element   A number in a matrix.

elimination   One method of solving linear systems; it involves adding or subtracting to rid a variable from one side of the linear system.

excluded value   A number that makes a rational expression undefined.

experimental probability   A probability based on repeated trials: the ratio of the number of successes to the total number of trials.

exponent   The number or variable that represents the number of times the base of a power is used.

extraneous solution   The solution of an equation that has been transformed and is also not the solution to the original equation.

factor   A whole number that divides another whole number with no remainder.

factorial   Written as $n!$, defined as $n! = n \times (n - 1) \times (n - 2 \times \ldots \times 3 \times 2 \times 1)$.

FOIL   A way to remember how to use the distributive property to multiply polynomials; it stands for First, Outside, Inside, Last.

fraction   A way to represent the quotient of two numbers.

function   A set of inputs and outputs where each input is paired with only one output.

greatest common factor   The greatest number that is a factor of two or more numbers.

histogram   A bar graph of data from a frequency table.

hyperbola   The graph of an inverse variation equation ($y = \dfrac{a}{x}$) or the graph of a rational function ($y = \dfrac{a}{x-b} + k$). The branches of the hyperbola approach but do not touch the axes.

identity property of addition   States that the sum of any number a and zero is a ($a + 0 = a$).

identity property of multiplication   States that the product of any number a and 1 is a ($a \times 1 = a$).

independent compound event   Two events where the first event has no impact on the second event. An example is flipping a coin and then flipping it again.

input   The number in the domain of a function.

inverse variation   The relationship between two variables so that $y = \dfrac{a}{x}$.

least common denominator   The product of the factors of the denominators with each common denominator used only once.

line of best fit   A line that models the trend on the data on a scatter plot.

linear equation   An equation which, when graphed, is a line.

linear function   The equation $Ax + By = C$ when $B \neq 0$.

linear system   Two or more linear equations with the same variables.

lower quartile   The median of the lower half of an ordered set of data.

matrix   An arrangement of numbers in rows and columns. For example, $\begin{bmatrix} 2 & -6 \\ 0 & 12 \\ 9 & 6 \end{bmatrix}$.

mean   The average of a set of numbers, $\bar{x} = \dfrac{x_1 + x_2 + \ldots x_n}{n}$.

mean absolute deviation   A measure of dispersion found with the formula $\bar{x} = \dfrac{|x_1 - \bar{x}| + |x_2 - \bar{x}| + \ldots |x_n - \bar{x}|}{n}$.

measures of dispersion   Describes the spread of data—range and mean absolute deviation are two examples.

median   The middle number of a set of numbers written in numerical order.

mode   The value that occurs most frequently in a set of numbers. There could be one mode, no mode, or more than one mode in a set of data.

monomial   A number, variable, or product of a number and more than one variable. For example, 103, $4x$, and $9m^3$.

numerator   The upper part of a fraction.

odds   The chance that something will happen. It can be favorable or unfavorable.

order of operations   The rules used to evaluate an expression involving more than one operation. Think "Please Excuse My Dear Aunt Sally" or "Purple Elephants Make Delicious Apple Scones."

output   A number in the range of a function.

overlapping events   Events that have at least one common outcome.

parabola   U-shaped graph of a quadratic function.

parallel lines   Two lines that do not intersect and have the same slope.

percent   A fraction with a denominator of 100.

permutation   An arrangement of objects where the order is important.

perpendicular line   Two lines that intersect at a right angle. The slopes of two perpendicular lines are negative reciprocals of each other.

polynomial   A monomial or sum of monomials.

principle   The original amount deposited into an account.

probability   The likelihood that an event will occur.

proportion   An equation where two ratios are equivalent; $\frac{a}{b} = \frac{c}{d}$ where $b \neq 0$ and $d \neq 0$.

quadratic equation   An equation that can be written as $ax^2 + bx + c = 0$ when $a \neq 0$.

radical equations   An equation with a radical expression that has a variable in the radicand.

radicand   The number or expression inside a radical symbol. The radicand of $\sqrt{12}$ is 12.

rate   A fraction that compares two quantities with different units. Miles per hour is an example of a rate.

ratio   A way to compare two numbers using division. The ratio of x and y can be written as x to y, $x:b$, or $\frac{x}{y}$.

rational equation   An equation that contains at least one rational expression.

rational expression   An expression that can be written as a ratio of two polynomials; the denominator cannot be 0. For example, $\frac{2x}{3x^2 - 7}$.

rationalizing the denominator   One method used to remove the radical from the denominator of an expression; done by multiplying the expression by 1.

reciprocal   Two numbers are reciprocal if their product is 1. When you divide by a number, it is the same as multiplying by the reciprocal.

scalar   A real number by which a matrix is multiplied.

scatter graph   A graph of data points used to determine a trend.

scientific notation   A number written as $x \times 10^n$ when $1 \leq x \leq 10$ and n is an integer. For example, 2.5 $\times 10^2$ or 7.8×10^{-8}.

simple interest   Interest paid only on the original deposit. Determined using the formula $I = prt$.

slope   The ratio of rise and run; represented by the variable m, equation for the slope is $m = \frac{y_2 - y_1}{x_2 - x_1}$.

slope-intercept equation   A linear equation written in the form $y = mx + b$ where m is the slope and b is the y-intercept.

square root   Represented by the symbol $\sqrt{}$.

square root function   A radical function where the equation has a square root.

standard form   A linear equation when written as $Ax + By = C$.

stem-and-leaf plot   A display of data that organizes the data by their digits.

subtraction property of equality   States that if the same number is subtracted from each side of an equation, the results are equal. If $a = b$ then $a - c = b - c$.

subtraction property of inequality   States that subtracting the same number from each side of an inequality does not change the relationship. If $a > b$ then $a - c$ is still $> b - c$.

system of inequalities   Two or more inequalities with the same variables.

trinomial   A polynomial with three terms, as in $4x^2 + 3x - 7$.

upper quartile   The median of the upper half of an ordered data set.

variable   A letter used to represent one or more numbers.

vertex   The highest or lowest point of a parabola.

vertical line test   Quick, easy way to determine if a graph represents a function. If a vertical line does not intersect more than one point, then it is a function.

x-intercept   The x-coordinate of a point where a graph crosses the x-axis.

y-intercept   The y-coordinate of a point where a graph crosses the y-axis.

zero product property   States that the product of any number and 0 is 0.

Index

H-I-J

K-L

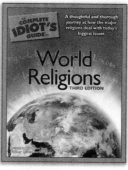

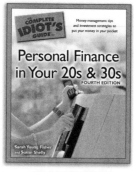

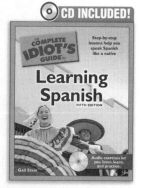